Science Diplomacy

New Day or False Dawn?

Science Diplomacy

New Day or False Dawn?

Editors

Lloyd S. Davis
University of Otago, New Zealand

Robert G. Patman
University of Otago, New Zealand

World Scientific

NEW JERSEY • LONDON • SINGAPORE • BEIJING • SHANGHAI • HONG KONG • TAIPEI • CHENNAI

Published by

World Scientific Publishing Co. Pte. Ltd.
5 Toh Tuck Link, Singapore 596224
USA office: 27 Warren Street, Suite 401-402, Hackensack, NJ 07601
UK office: 57 Shelton Street, Covent Garden, London WC2H 9HE

Library of Congress Cataloging-in-Publication Data
Science diplomacy : new day or false dawn / editors, Lloyd S. Davis, University of Otago, New Zealand, Robert G. Patman, University of Otago, New Zealand.
pages cm
Includes index.
ISBN 9789814440066 (hardcover : alk. paper) -- ISBN 981444006X (hardcover : alk. paper)
1. Science--Political aspects. 2. Science--International cooperation. 3. Globalization.
I. Davis, Lloyd Spencer, 1954– editor. II. Patman, Robert G., editor.
Q175.5.S354 2014
303.48'3--dc23

2014021379

British Library Cataloguing-in-Publication Data
A catalogue record for this book is available from the British Library.

In-house Editor: Philly Lim

Typeset by Stallion Press
Email: enquiries@stallionpress.com

Printed in Singapore

ACKNOWLEDGMENTS

We owe a debt of gratitude to a number of people and institutions for assistance in the preparation of this book. The concept for this volume evolved from the occasion of the 46th University of Otago Foreign Policy School. As Co-Directors of that School and editors of this book, we wish to acknowledge the substantial support that made this volume possible.

First, we would like to thank our colleagues on the Academic Committee of the 46th School: Ms Jan Brosnahan, the School Coordinator, Associate Professor Jenny Bryant-Tokalau, Associate Professor Paul Hansen, Mr Pierce Lane, Associate Professor Jacqueline Leckie, Mr Eliot Lynch, Mrs Betty Mason-Parker, Professor Philip Nel, Dr Maria Pozza, Dr Chris Rosin, and Dr Paola Voci.

Second, the research assistance of Ms Sarah Macindoe, a Master of International Studies (MIntSt) graduate at the University of Otago, in the preparation of this volume deserves a special mention. Sarah did an outstanding job in supporting the editors; helping to facilitate coordination and communication with the contributors to the book; writing a chapter of her own; and contributing to the multi-authored chapter, which sets the scene for this volume.

Third, we would like to thank all the contributors to this book. They constitute a distinguished international team of specialists. They cheerfully accepted our editorial guidelines and took the time to revise their initial drafts into polished and stimulating chapters.

Fourth, we wish to express our sincere thanks for the encouraging and patient support for the project given by the staff at World Scientific Publishing. That assistance was greatly appreciated by the editors.

Fifth, we wish to thank all those organisations whose support and sponsorship played a significant role in bringing together our team of contributors. Without this support, it would have been difficult to develop the idea of this book. We are grateful to: the Australian High Commission, Wellington; the New Zealand Ministry of Foreign Affairs and Trade, Wellington; the

New Zealand Ministry of Foreign Affairs and Trade's Seriously Asia Programme; the New Zealand Ministry of Research, Science and Technology, Wellington; Orbit Corporate Travel, Dunedin; and the University of Otago.

Finally, and most importantly, we should especially like to thank our families, particularly our partners, Wiebke and Martha. Throughout the duration of this book, they were supportive in every possible way.

Lloyd S. Davis and Robert G. Patman
University of Otago,
New Zealand.
28 April 2014

About the Editors

Lloyd S. Davis is an internationally recognized scientist, as well as an award-winning author and filmmaker. Currently the inaugural Stuart Professor of Science Communication at the University of Otago, he has authored over 130 refereed scientific publications. He is regarded as a world authority on penguins and academic honours bestowed upon him have included a Fulbright Fellowship, an Anzac Fellowship and a Prince and Princess of Wales Science Award. He is the author of seven books and has edited two more. These include *The Plight of the Penguin*, winner of the New Zealand Children's Book of the Year Award and *Looking for Darwin*, winner of the CLL Writers Award for Nonfiction. His eighth book, *Professor Penguin*, is due to be published in late 2014.

Robert G. Patman's research interests concern US foreign policy, international relations, global security, great powers and the Horn of Africa. He is an editor for the journal *International Studies Perspectives*, and the author or editor of 11 books. Recent publications include a volume called *Strategic Shortfall: The 'Somalia Syndrome'* and the *March to 9/11* (Praeger, 2010) and two co-edited books titled *The Bush Leadership, the Power of Ideas, and the War on Terror* (Ashgate, 2012), and *China and the International System: Becoming a World Power* (Routledge, 2013). He is a Fulbright Senior Scholar, a Senior Fellow at the Centre of Strategic Studies, Wellington, an Honorary Professor of the NZ Defence Command and Staff College, Trentham, and provides regular contributions to the national and international media on global issues and events.

About the Contributors

Edison Liu was in Hong Kong in 1952. Edison obtained his B.S. in chemistry and psychology, as well as his M.D., at Stanford University. He served his internship and residency at Washington University's Barnes Hospital in St. Louis, followed by an oncology fellowship at Stanford. From 1982 to 1987 he was at the University of California, San Francisco, first in a haematology fellowship at Moffitt Hospital and then as a postdoctoral fellow in the laboratory of Nobel Laureate J. Michael Bishop, while also serving as an instructor in the School of Medicine.

From 1987 to 1996 he was an assistant professor at the University of North Carolina at Chapel Hill in medicine and oncology at the School of Medicine, and rose to full professor directing the UNC Lineberger Comprehensive Cancer Centre's Specialized Program of Research Excellence in Breast Cancer. He was also the director of the Laboratory of Molecular Epidemiology at UNC's School of Public Health, chief of medical genetics, and chair of the Correlative Science Committee of the national cooperative clinical trials group, CALGB. In 1997, he was appointed the Scientific Director of the Division of Clinical Sciences at the National Cancer Institute (USA) overseeing the intramural clinical sciences program for the institute. Between 2001–2011, Dr. Liu was the founding Executive Director for the Genome Institute of Singapore. He joined the Jackson Laboratory in 2012 as its President and CEO, and as the Director of the Jackson Laboratory NCI Designated Cancer Centre.

Vaughan Turekian is the Chief International Officer for the American Association for the Advancement of Science (AAAS) and the Founding Director of AAAS's Center for Science Diplomacy. In 2011, Vaughan became the founding Editor-in-Chief of *Science and Diplomacy Quarterly*. Vaughan was formerly an AAAS Diplomacy Fellow at the US Department of State in the Climate Office, before serving as the Special Adviser to the Under Secretary of State for Global Affairs. He is the two-time recipient of the Department's Superior Honour Award for his work on climate change and avian influenza. Prior to his time at the State Department, he worked at the National Academy of Sciences (NAS). In 2001, he was the Study Director for the NAS report on climate change science requested by the White House. He is currently an adjunct professor in Georgetown's School of Foreign Service and a Distinguished Visiting Scholar at University College London. Vaughan received his masters and doctorate in atmospheric geochemistry from the University of Virginia. He is a graduate of Yale University with degrees in geology and geophysics, and international studies.

Sefton Darby is the National Manager for Minerals at the Ministry of Business, Innovation and Employment (MBIE) in New Zealand. Prior to taking up his current role he worked for the Newmont Mining Corporation, the World Bank, the UK Government (Cabinet Office and the Department for International Development) and ran his own international development consultancy with various government, multilateral and NGO clients. During this time he worked on resource governance issues in a number of countries in West Africa, the former Soviet states, and the Asia-Pacific. He is a graduate of the universities of Otago and St Andrews. His contributed work in this book is entirely based on his previous experience in international aid and development and reflects his personal views only — in no way does it reflect the views of MBIE.

Gary Wilson is Director of the New Zealand Antarctic Research Institute and Professor of Marine Science at the University of Otago and also holds an adjunct position in the Geology Department at Otago. Gary has participated in and led more than 20 expeditions to the Antarctic and Subantarctic supported by the New Zealand and US Antarctic Programmes. He has contributed to more than 80 scientific papers in peer research journals. He has held the Byrd Fellowship at the Ohio State University, the Blaustein Visiting Professorship at Stanford University and in 2006 he received a Sir Peter Blake Leadership award for his role in bringing together the multinational team for the ANDRILL project, which he chaired between 2004 and 2009.

A University of Auckland Distinguished Professor, **Peter Gluckman** is Professor of Paediatric and Perinatal Biology. He was formerly Chairman of the Department of Paediatrics and Dean of the Faculty of Medical and Health Sciences for nine years. He has been extensively involved in many aspects of science, health and educational policy development. He is currently Chief Science Advisor to the Prime Minister of New Zealand, a post he held since July 2009. In 2009, he became a Knight of the New Zealand Order of Merit replacing the 2008 Distinguished Companion of the New Zealand Order of Merit, for services to medicine and having previously been made a Companion of the Order in 1997. In 2001, he received New Zealand's top science award, the Rutherford Medal. Sir Peter is an international advocate for science, promoting the translation of discoveries in biomedical research into improvements in long term health outcomes. His work with organisations such as the WHO has brought growing recognition of the importance of a healthy start to life. He is the author of over 600 scientific papers and reviews and editor of eight books, including three influential textbooks in his subject area.

Joan Leach (BA Hons, BSc, MA, PhD) convenes the Science Communication Program at the University of Queensland and is Associate Professor of Rhetoric in the School of English, Media Studies and Art History. Her research is centred on public engagement with science, medicine and technology and she has been active in the Australian government's recent initiatives toward "Inspiring Australia". She is currently researching the role of popular science in the globalisation of science since the 1960s, a project funded by the Australian Research Council. She has published extensively about science communication, including a 2012 book *Rhetorical Questions of Health and Medicine*, and was editor of the *International Journal, Social Epistemology* from 1997 to 2010 and is now an executive editor for the journal. She held academic posts at the University of Pittsburgh (USA) and Imperial College London before moving to Brisbane in 2005. While remaining transfixed by science, she advocates for better science communication which critically examines the social impact of science, technology and biomedicine.

Stephen Goldson's scientific career has been as an entomologist working on New Zealand's very damaging exotic pasture pest species (especially weevils). Through this work he was able to make a useful contribution to the suppression of these pests based on imported parasitoid wasp biological control agents. He is now a principal scientist in AgResearch Ltd., New Zealand and Theme Leader in the Bio-Protection Research Centre at Lincoln University. In the mid-1990s, Stephen worked part-time as Science Adviser to the then Minister of Research, Science and Technology. A similar arrangement has been underway since 2009 whereby he has a part-time appointment as strategist to the Chief Science Advisor to the New Zealand Prime Minister, Professor Sir Peter Gluckman. On the basis of his research contribution he was elected as a Fellow of the New Zealand Institute of Agricultural and Horticultural Science (1998) and the Royal Society of New Zealand (2006).

In 2014 he was appointed as an Officer of the New Zealand Order of Merit for his service to science in New Zealand.

Jeffrey H. Boutwell is former Executive Director of the Pugwash Conferences on Science and World Affairs, an international scientific organisation founded in 1957 to seek ways of eliminating nuclear weapons and other weapons of mass destruction, which in 1995 received the Nobel Peace Prize. He is the author of *The German Nuclear Dilemma* (Cornell University Press, 1990) and has written widely on nuclear weapons issues, small arms and light weapons, environmental degradation and conflict, and Israeli-Palestinian security issues. He received his PhD from the Massachusetts Institute of Technology and his MSc from the London School of Economics. Previously, he was a staff aide on the National Security Council during the Carter administration, Associate Executive Officer at the American Academy of Arts and Sciences in Boston, and an editor and reporter at the City News Bureau of Chicago.

Manjana Milkoreit is a Postdoctoral Research Fellow with the Walton Sustainability Solutions Initiatives at ASU's Julie Ann Wrigley Global Institute of Sustainability and an affiliate of the Consortium for Science, Policy and Outcomes. Her research concerns the role of cognition and emotion in climate change politics and sustainability decision making. She is particularly interested in the use of scientific knowledge in political, policy and governance processes, the role of ideologies in advancing or preventing effective societal responses to climate change, and the cognitive ability to imagine distant futures. Manjana's previous work analysed whether and how cognitive processes influence the search for cooperative multilateral solutions in the UNFCCC. She has also studied the changing role of the emerging powers (China, India, Brazil, and South Africa) in the global climate negotiations. She is an active member of international, multidisciplinary research networks on complexity, resilience and social-ecological systems, including the Resilience Alliance and the Waterloo Institute for Complexity and Innovation.

Manjana holds a Master in Public Policy from the Harvard Kennedy School and a PhD in Global Governance from the University of Waterloo.

Daryl Copeland is an educator, analyst, consultant and author of *Guerrilla Diplomacy: Rethinking International Relations* (Lynne Rienner Publishers, 2009), and has written over 100 articles and book chapters for the scholarly and popular media. A former Canadian diplomat who served in Thailand, Ethiopia, New Zealand, and Malaysia, he is now Senior Fellow at the Canadian Defence and Foreign Affairs Institute, and specialises in the relationship between science, technology, diplomacy and international policy. He holds teaching appointments at the Diplomatic Academy of Vienna, as well as Ottawa, Otago (NZ), and East Anglia Universities. Daryl is a peer reviewer for the University of Toronto Press, *Canadian Foreign Policy, the International Journal*, and *The Hague Journal of Diplomacy*, and is a member of the Editorial Board of the journal *Place Branding and Public Diplomacy* and for the Advisory Board of Canadian Foreign Policy. In 2009, he was Research Fellow at the Centre on Public Diplomacy, University of Southern California and from 2009 to 2011 he served as Adjunct Professor at the Munk School of Global Affairs. In 2000, he received the Canadian Foreign Service Officer Award; and in 2012, he received the Molot Prize for foreign policy analysis. For more information and commentary, see www.guerrilladiplomacy.com, and follow him on Twitter @GuerrillaDiplo.

Sarah Macindoe holds a BA (Hons) in Politics and a BSc in Genetics from the University of Otago in New Zealand, and in February 2014 was awarded a Master of International Studies (Distinction) by the same institution. Sarah's studies prompted an interest in issues surrounding resource scarcity — particularly in the area of food security — and in international efforts to combat and manage such threats. Sarah now works as an analyst in the National Assessments Bureau, a unit of the Department of Prime Minister and Cabinet in Wellington, New Zealand. The views expressed in Sarah's chapter are her own and do not necessarily reflect the views of the New Zealand government.

Cathleen A. Campbell joined CRDF Global in 2002 and was appointed President and CEO in 2006. Cathleen brings three decades of international science, technology and innovation policy and programme management experience. She has worked with countries in Eurasia, the Middle East, Asia and Latin America. Cathleen served from 1998 to 2002 as director of the Office of International Technology Policy and Programs, Department of Commerce. From 1995 to 1997, she was a senior policy analyst in the White House Office of Science and Technology Policy. Cathleen was the US State Department's program officer for Soviet/Russian science and technology affairs from 1989 to 1994. Previously she held research positions in the private sector. Cathleen serves on the External Advisory Board, Pennsylvania State University's School of International Affairs; the US-Russia Innovation Working Group; the Advisory Board, Muslim-Science.com; and the Innovation Index Advisory Committee, US-Israel Science and Technology Foundation.

CRDF Global is an independent non-profit organization that promotes international science and technology cooperation through grants, technical resources, training and services. With a staff of 150 located in five countries, CRDF Global implements high-impact programs and services in over forty countries.

Atsushi Sunami is Professor at the National Graduate Institute for Policy Studies and a former special adviser to the Minister of State for Science and Technology Policy, the Cabinet Office of Japan.

Tomoko Hamachi is professional staff for the National Graduate Institute for Policy Studies Innovation, Science and Technology Policy Program, Japan.

Shigeru Kitaba is fellow and manager at the Department of Strategy, Centre for Research and Development Strategy, Japan Science and Technology Agency.

Contents

Introduction

CHAPTER 1

The Emergence of Science Diplomacy

Vaughan C. Turekian with Sarah Macindoe, Daryl Copeland, Lloyd S. Davis, Robert G. Patman, and Maria Pozza

Introduction

Major structural changes in the international system over the last three and half decades have raised a big question mark over the Westphalian principle of state sovereignty that assumes that a state — subject to international recognition — exercises legal, unqualified and exclusive control over a designated territory and population. The end of the Cold War in the late 1980s and the process of deepening globalization served to profoundly alter the global political context. These changes seem to make the world a much smaller and more interconnected place, but one that is seemingly fragmented by the erosion of the autonomy of the sovereign state and the rise of intra-state conflict. These changes seem to make the world a much smaller and more interconnected place, but one that is seemingly fragmented by the erosion of the autonomy of the sovereign state and the rise of intra-state conflict. In this new environment where shared challenges — such as food security, water availability, health management — require strong interactions between the science and technical communities across borders, science has taken on a role of greater importance in the international system. As a consequence, a globalizing world has eroded the old dichotomy between science and diplomacy, and helped to facilitate the emergence of science diplomacy whereby scientific collaborations among nations are necessary to tackle increasingly common challenges.

In this introductory chapter, we explore the evolving relationship between science and diplomacy. The chapter proceeds in five stages. The first section delineates the concept of Science diplomacy. The second considers the historical association between science cooperation and international relations. In the third part, we examine the international circumstances that have contributed to the rise of Science diplomacy. The fourth section identifies three types of activity that are related to Science diplomacy and uses these categories as an analytical framework for organising the discussion in this book. Finally, the concluding section provides a rationale for this volume and outlines the essays that comprise it.

The Concept of Science Diplomacy

The term 'science diplomacy' is a relatively new one and reflects the fusion of two previously distinct elements. *Science* is an evidence-based form of knowledge acquisition. It is founded upon empirical methods of experimentation and the repeated verification of results. Science is neither inherently political nor ideological, but represents a type of universal language, a vector of transnational communications that poses fundamental questions about the nature of things. The scientific ethos of objective experimentation through trial and error has broad appeal: it promotes merit (through peer review); openness (through publication); and civic values and citizen empowerment (through the encouragement of respect for diverse perspectives). In a public opinion survey reported in New Zealand on 20 June 2011, scientists were identified as the most trusted people in the country, and science as the most respected profession (TVNZ, 2011).

Diplomacy is a non-violent approach to the management of international relations characterized by dialogue, negotiation and compromise, often by a country's representatives abroad, and involves the art of dealing with people or their representatives in a sensitive and tactful way. Diplomats pursue and deliver international policy objectives on behalf of governments, and it is that connection to the state which sets diplomatic practice apart from the international lobbying, advocacy and public relations activities engaged in by business and civil society actors.

Science diplomacy, therefore, is the process by which states represent themselves and their interests in the international arena when it comes to areas of knowledge — their acquisition, utilization and communication — acquired by the scientific method. It is a crucial, if under-utilized, specialty within the diplomatic constellation that can be used to address global issues, enhance co-operation between countries and leverage one country's influence over another.[1] In this regard, Science diplomacy is a significant generator of *soft power* (Nye, 2004) — that potent form of attraction that harnesses national image, reputation, and brand. More broadly, science diplomacy is an effective emissary of essential values such as evidence-based learning, openness and sharing. Science diplomacy is increasingly critical to addressing many of the planet's most urgent challenges — such as management of the global commons, faltering public health systems, and the threat of collapsing ecosystems. It can also be used to enhance one nation's interests with respect to another or to defuse international tensions.

Science diplomacy's direct relationship with national interests and objectives distinguishes it from other forms of international scientific co-operation, which are sometimes commercially oriented and often occur without direct state participation. International scientific co-operation motivated by advancing science and is typically a win-win proposition, with private sector or civil society partners collaborating to produce, for example, better medications, cleaner water, improved hygiene or more disease-resistant crops. All parties reap the rewards. Science diplomacy is also founded upon mutuality and common cause, with the relationship being a central motivator for the cooperation. However, because national interests and the state are often implicated, motives may diverge and the outcomes may be asymmetrical, particularly if there are negotiations involved. A whole constellation of international scientific programs and exchanges undertaken during the second half of the last century come to mind by way of illustration, as do contemporary international discussions on issues such as the terms and conditions of resource access or environmental protection. While science itself may be apolitical, research and development in areas of Science and Technology (S&T) is often highly politicised, with countries keeping a firm eye on their scientific investments and on any potentially lucrative results. As Perkins argues, growing competitiveness — especially surrounding patents for drugs and new plant and animal varieties or the development of renewable energy

---

[1]A useful synopsis is offered in *New Frontiers in Science Diplomacy* (The Royal Society, 2010). This publication sets out three distinct activity areas within science diplomacy: informing foreign policy objectives with scientific advice (science in diplomacy); facilitating international science cooperation (diplomacy for science); using science cooperation to improve international relations between countries (science for diplomacy).

sources — suggests that 'tensions between national commercial interests and ambitions for goodwill between nations may not be easy to reconcile' (Perkins, 2012). For example, the Centre for Global Development reports that the commitment to development on the part of the US and a number of members of the EU — particularly Germany and Sweden — is regularly undermined by attempts to '[restrict] the flow of innovations to developing countries' by incorporating 'TRIPS-Plus'[2] measures into bilateral free-trade agreements. US trade negotiators have pressured developing countries to agree not to force immediate licensing of patents even if this would serve a compelling public interest, such as with HIV/AIDS drugs (Barder and Krylova, 2013).

Thus, international science cooperation and science diplomacy are overlapping endeavours: they are related, yet analytically separate. International science cooperation is mainly concerned with the advancement of scientific discovery *per se*, while the central purpose of science diplomacy is often to use science to promote a state's foreign policy goals or inter-state interests. In other words, international science cooperation tends to be driven by individuals and groups, whereas science diplomacy, while it may derive from the efforts of individuals, often involves a state-led initiative in the area of scientific collaboration. International science cooperation, therefore, may or may not encompass science diplomacy.

Conceptually, the idea of science diplomacy seems to be characterised by a potential tension in the relationship between the two key paradigms that comprise it. Diplomacy traditionally requires that practitioners have a good general knowledge of concerns relating to state interests, but diplomats would not typically see their professional remit extending to the details and complexities of modern science. At the same time, science involves the search

[2]The 1995 Agreement on Trade-Related Aspects of Intellectual Property Rights (TRIPS Agreement) is to date the most comprehensive multilateral agreement on intellectual property (IP), and provides a prescriptive regime for the protection and enforcement of intellectual property rights. With the remarkable upsurge in the number of free-trade agreements being signed in the past decade, the 'post-TRIPS' era has seen efforts to strengthen the protections for IP beyond those established under TRIPS, creating the 'TRIPS-Plus' phenomenon. Developing countries in particular have come under increasing pressure to enact these tougher 'TRIPS-Plus' provisions in their patent laws. Bilateral science and research and development cooperation agreements constitute an indirect form of 'TRIPS-Plus'.

for verifiable knowledge and this process, almost by definition, is a collaborative activity and one which is likely to straddle national boundaries. However, scientific practitioners are not always sensitive to the diplomatic implications of their research-led cooperation across state boundaries. In the words of a 2010 joint publication by the Royal Society and the American Association for the Advancement of Science (AAAS), 'scientists and diplomats are not obvious bedfellows. While science is in the business of establishing truth, Sir Henry Wotton, a 17th century diplomat, famously defined an ambassador as "an honest man sent to lie abroad for the good of his country"' (The Royal Society, 2010: 1).

The link between international science cooperation and international relations

Notwithstanding divergent orientations, there is a long historical association between science and international cooperation. The post of Foreign Secretary of the Royal Society was instituted in 1723, nearly six decades before the British Government first appointed a secretary of state for foreign affairs, and in 1941 Sir Charles Galton Darwin FRS (the grandson of Charles Darwin) was appointed Director of the Central Scientific Office in Washington, becoming the UK's first accredited scientific representative abroad. Just one year later, Joseph Needham FRS was made Head of the British Scientific Mission in China. He actively promoted the formation of an 'International Science Co-operation Service', and his lobbying led to the inclusion of natural sciences within the mandate of the United Nations Educational, Scientific and Cultural Organization (UNESCO) (The Royal Society, 2010: 1).

The United States also has a long history of involvement in cooperative international scientific efforts. In the early 1970s, as the country was winding down its involvement in a controversial war in Southeast Asia that clearly demonstrated the limits of US hard power, an adviser to then-Secretary of State Henry Kissinger stated to *Science* magazine that '[the Secretary of State] thinks that America's ability to contribute money and run the world in the old fashioned way ... is now over. What we can contribute — and what the world wants — is our technological capabilities' (Wade, 1974).[3]

---

[3]Similarly, in an address to the United Nations session on development in April 1974,

This focus on the role of science and technology became a central element of US foreign policy outreach to allies and adversaries alike during the course of the Cold War. In a 1985 address to the nation just days before meeting with Soviet leader Mikhail Gorbachev for the first time, President Ronald Reagan stated: 'We can find, as yet undiscovered, avenues where American and Soviet citizens can cooperate fruitfully for the benefit of mankind… . In science and technology, we could launch new joint space ventures and establish joint medical research projects' (Regan, 1985). Two years later, John Negroponte, the President's Assistant Secretary of State for Oceans and International Environmental and Scientific Affairs (OES), further articulated the Administration's view during congressional testimony: 'It would be shortsighted of us not to recognize that it is in our national interest to seek to expand scientific cooperation with the Soviet Union'.[4] In many ways, the Cold War period initiated the beginnings of science diplomacy, as states used scientific collaboration to build bridges and connections despite the existence of great political tensions.

While perhaps the interactions between the United States and the USSR provide the most well-known historical case of linking scientific cooperation to foreign relations, they are by no means the only example. Throughout the second half of the 20th century, science played many important roles in diplomacy. At a White House state dinner for Japanese Prime Minister Hayato Ikeda in 1961, President John Kennedy made US diplomatic history by announcing the US-Japan Committee on Science Cooperation, the first of its kind. Kennedy had followed the advice of his Ambassador to Japan — the illustrious scholar and Harvard professor Edwin Reischauer — and created the committee as part of a broad effort to repair 'the broken dialogue' between the intellectual communities of the two countries.[5] The National

---

Kissinger noted that we 'now apply science to the problems which science has helped to create', pinpointing agricultural technology, birth control, weather modification and energy as areas of particular interest.

[4]As quoted in Turekin and Neureiter (2012).

[5]In the wake of rising tensions between Tokyo and Washington over the revised Security Treaty between the two countries, Reischauer wrote an article for *Foreign Affairs* in which he pointed to the 'weakness of communication between the Western democracies and opposition elements in Japan' (see Reischauer, 1960). His article so impressed President-elect John F. Kennedy that he was subsequently appointed United States Ambassador to Japan.

Science Foundation's implementation of that cooperative science program has endured for over half a century, evolving with the times and delivering great benefits to both countries.

International scientific cooperation, while strongly linked to the Cold War experience of the United States, also served as an important instrument in the wider global context. For example, after World War II had divided the European continent, collaboration on scientific endeavours served as a significant ingredient in efforts to improve inter-state relations. In 1954, CERN, the European Organisation for Nuclear Research, was established. It was a major project in which the Federal Republic of Germany (FRG) was able to work with former European adversaries such as France. According to a former CERN director, Horst Wenninger, in the aftermath of the Second World, 'cooperation between [European] nations was simpler in science than in other fields' (Prolavorio, 2013) and helped play a part in Franco-German rapprochement that culminated in the Elysée Treaty of January 1963.

Similarly, scientific interactions between Israel's Weizmann Institute and Germany's Max Planck Society provided a channel for the first high-level discussions between the countries after World War II. Originating in the late 1950s, collaboration between the two bodies precipitated a historic agreement in 1964 which facilitated the transfer of German government funds to Weizmann Institute research projects, hereby fostering a wide range of scientific exchanges between the Institute, the Max Planck Society, and other German universities. Such ties 'helped lay the foundation not only for German-Israeli scientific cooperation, but also for the establishment of diplomatic relations between the two countries one year later' (Weizmann Institute of Science, 2013). In January 2012, the two groups announced the creation of a joint Centre for Archaeology and Anthropology, marking their more than five decades of scientific partnership. It is hoped that the Centre will not only strengthen ties between the Max Planck Society and the Weizmann Institute, but may encourage an expansion of scientific ties between Israel and its regional neighbours. The Institute's Professor Stephen Weiner has expressed enthusiasm regarding the potential political and diplomatic benefits of the initiative, noting that 'just as happened in relations with Germany, now too scientific collaboration could have a broader impact, helping to promote peaceful ties in the Middle East' (Weizmann Institute of Science, 2013).

More recently during the post-Cold War era, science outreach has provided an important — and often first — step in the EU's efforts to expand its diplomatic footprint into such places as post-communist East Europe and an Islamic country like Turkey. Such efforts are also increasingly taking place in other parts of the world. Science cooperation has a powerful role in helping countries as they look to build stronger regional partnerships. Within the East African Community (Burundi, Kenya, Rwanda, Tanzania, and Uganda), a presidential-level initiative to better align and integrate this diverse, populous and historically unstable region is drawing on the promise of scientific cooperation.[6] By sharing costs and resources and increasing the interaction of students and researchers, such technical cooperation can help the region increase its prosperity while contributing to more sustainable regional links. However, it must be acknowledged that not all science diplomacy has been devoted to civilian and diplomatic purposes. A particular area of concern — exemplified by the A. Q. Khan network — has been covert collaboration in the field of nuclear weapons technology and the attendant risk of nuclear proliferation.

On a more general note, scientific advice is crucial for diplomats, and growing recognition of this need has resulted in the rise of international scientific advisory bodies since the 1950s (National Research Council, 2002: 6). In 1957 and 1958 a global community of scientists joined together in a sharing of information and research, naming the period the International Geophysical Year (IGY). The International Council of Scientific Unions arranged for much scientific collaboration across borders, and to some extent did not recognise state borders at all. Sixty-seven states participated in the IGY by prior international agreements settled by the negotiation of diplomats. Upon the success of the IGY collaboration, other scientific research programmes arose which have led to institutions such as the Scientific Committee on Antarctic Research.

While the role of international cooperation in science has a long history, the interaction between science and the conduct of a nation's foreign policy

---

[6]Growing recognition of the centrality of science, technology and innovation to economic growth, development and regional integration has led the members of the East African Community (EAC) to adopt a number of programmatic initiatives and protocols, such as the Protocol on Science, Technology and Innovation, to foster broader cooperation in this area. For more information see Tumushabe and Omar-Mugabe (2012).

does not have such an intertwined past. However, in recent years there has been an increased focus on issues at the interface of science and foreign policy, leading to greater emphasis on the relationship between science and diplomacy.

Globalization and the Rise of Science Diplomacy

Structural changes in the mid to late 1980s began to challenge a compartmentalised, state-centred understanding of global politics. The aftershocks of the end of the Cold War and intensified globalization were associated with the growth of international linkages and a reduction in the capacity of nation-states to act independently. The time of absolute and exclusive national sovereignty began to wane as the traditional boundaries between domestic and external policy roles of the sovereign state were blurred by the impact of globalization (Scholte, 2001: 14). The latter could be broadly defined as the intensification of technologically driven links between societies, institutions, cultures and individuals on a worldwide basis. Above all, it was revolutionary changes in communication and information technologies in the 1980s — advances in personal computing and the development of the internet — that effected a compression of time and space by reducing the time taken to cross geographical boundaries. This process facilitated 'networks of interdependence at multicontinental distances' (Keohane and Nye, 2000). As an upshot, the world began to be perceived as a smaller place, with issues relating to the environment, economics, politics and security intersecting more deeply an at more points than was previously the case (Clark, 1997: 15).

The advent of globalization initiated a debate over the role of the sovereign state in the modern world.[7] Three rival schools of thought can be identified. The *hyperglobalists* contend that the growing interconnectedness of states through globalization gradually negates the significance of territorial boundaries and paves the way for the decline of the sovereign nation-state (Held and Anthony, 1999: 4). In contrast, the *realists* or *skeptics* basically believe that little has changed in the international arena. The skeptics argue

[7]The seminal text on the various approaches to globalisation is Held *et al.* (1999). In particular, see the Introduction (pp. 32–86).

that the impact of globalization on the sovereign state is much exaggerated. From this viewpoint, the state is not the victim of this process, but its main architect (Held and Anthony, 1999: 8).

On the other hand, the *transformationalists* reject the tendency of both the hyperglobalists and the skeptics to juxtapose state sovereignty and globalization. For transformationalists, state sovereignty is a dynamic concept that is simply undergoing a new phase in its evolution as states respond to the costs and the benefits of the globalization process. This environment is not only widening the opportunities for many states to interact diplomatically, but is also obliging states to recognize that many diplomatic challenges they are now facing are complex and can only be resolved through multilateral or international action. In this era of globalisation, the most profound challenges to human survival — climate change, diminishing bio-diversity, public health, food insecurity and resource scarcity, to name but a few — are rooted in science and driven by technology. Thus, according to the transformationalists, globalization is a 'mega trend' that is not only changing the nature of the sovereign state but is also providing the impetus for the rise of science diplomacy.

Growing interest in science diplomacy is, therefore, accompanying an evolution in international relations, and is in some ways a function of such global change. As a more disaggregated diplomatic system — consisting of dynamic networks of lawyers, scientific bodies, non-governmental organisations and the media — takes shape, (The Royal Society, 2010: 3) 'track II' diplomacy involving scientists, science and technology based business groups, and scientific regulatory advisors is acquiring a heightened significance. While science has always transcended borders, the growing ease with which such 'track II' initiatives can be accomplished is in large part due to the unprecedented mobility of ideas, people and information that characterises the globalisation age (National Academy of Sciences, National Academy of Engineering, and Institute of Medicine, 2011: 7–9).

The Parameters of Science Diplomacy

The convergence of two words — science and diplomacy — has produced an umbrella term that according to the British Royal Society and the American

Association for the Advancement of Science (AAAS) encompasses at least three main types of activities:

- *Diplomacy for Science*
- *Science in Diplomacy*
- *Science for diplomacy*

Diplomacy for science

One dimension of science diplomacy — diplomacy *for* science — seeks to 'facilitate international cooperation, whether in pursuit of top-down strategic priorities for research or bottom-up collaboration between individual scientists and researchers' (The Royal Society, 2010). While scientists and diplomats typically come from different backgrounds and experience very different training, there are many areas where their interaction is central to advancing the scientific enterprise.

For instance, while the science and technology community has had great interest in developing large scale and deployable fusion energy as a way to produce cheap, clean and abundant energy, the technical challenges have been formidable, as have the costs. As a result, there has been great interest within the physics community in developing large-scale multinational experimental platforms that could support such advanced science without decimating budgets. Working at the multinational level (at first involving China, Europe, Japan, the Republic of Korea, Russia and the United States), the international science community began to plan for such an international project. As Harding *et al.* (2012) noted, laying the diplomatic foundations was as important as overcoming the technical challenges:

> In addition to design and cost, there was no agreement on a legal and policy structure that would be appropriate for creating and sustaining an international facility and experiment. New approaches were needed for a form of agreement and organization that would allow partners with diverse political and legal systems to work together on a science experiment of this magnitude.

The need for cooperation between the diplomatic and scientific communities on such large multilateral programs is the principal driver behind diplomacy for science. This second dimension of science diplomacy has played a crucial role in enabling many other international scientific initiatives — such as the International Space Station, the Square Kilometer Array (SKA) project, the International Thermonuclear Experimental Reactor (ITER)[8] and the SESAME synchrotron[9] — to get off the ground. Diplomacy is therefore a key facilitator of science and technology research and development, allowing for communication and collaboration across and beyond national borders.

The work undertaken by the SKA project members across Australia and South Africa also illustrates the importance of diplomacy for science. The projected cost of the project will need to be met by the contributions of participating countries. Similarly, the project has been expanded to two locations in order to utilise the benefits both sites have to offer. Australia has superior radio silence and facilities well suited for low frequency research, whereas South Africa geographically is the ideal candidate for medium and high frequency analysis. Flagship international initiatives such as the Large Hadron Collider (LHC) also rely on the effective utilisation of diplomacy for science.[10] These projects carry enormous costs and risks, but are increasingly vital in areas of science that require large upfront investments in infrastructure, which are beyond the budget of any one participating country. In this sense, international scientific projects require diplomatic input.

---

[8]A fusion experimental research facility was first proposed after the standoff over nuclear disarmament at the Reykjavik Summit in October 1986, with collective design efforts beginning in 1988. The final ITER Agreement, signed in November 2006, emphasises the potential for diplomacy for science to enable large-scale, capital-intensive international projects. For more information see (Harding *et al.*, 2012).

[9]SESAME, or Synchrotron-light for Experimental Science and Applications in the Middle East, is a major intergovernmental scientific facility hosted by Jordan designed to 'foster scientific and technological capacities and excellence in the Middle East and the Mediterranean region' and 'build scientific links and foster better understanding and a culture of peace through scientific collaboration'. Skilful diplomacy and international cooperation have been central to the early stages of the project's implementation. For more information see Smith (2012).

[10]The Large Hadron Collider (LHC) housed at CERN in Switzerland is the most powerful particle accelerator ever built, and allows scientists to reproduce the conditions that existed within a billionth of a second after the Big Bang. 27km long and weighing more than 38,000 tonnes, the LHC is the product of a collaborative effort by 20 countries and an enormous international community of scientists and engineers working in multinational teams both at CERN and around the world. See Science and Technology Facilities Council (STFC) (n.d).

But such expansive multinational efforts are only the tip of the iceberg. Bottom-up collaboration takes place daily between institutions and individual scientists, and the strengthening of personal and professional relationships at this level is proving instrumental in driving crucial science and technology research and innovation. As the AAAS and Royal Society argue, 'the stereotype of the scientist as a lone genius no longer holds true. The scientific enterprise is now premised on the need to collaborate and connect' (The Royal Society, 2010: 6). Globally, we are increasingly seeing the emergence of 'an invisible college of researchers who collaborate not because they are told to but because they want to ... because they can offer each other complementary insight, knowledge or skills' (Wagner, 2008).

Science in diplomacy

Many of the major challenges facing states are increasingly global in nature and scale, and have science and/or technology in the fingerprint of their cause or cure. Science *in* diplomacy describes the role of science — and technology — in providing advice to inform and support foreign policy objectives. The function of science in diplomacy should be to ensure the effective uptake of high quality scientific advice by policymakers (National Research Council, 2002). The scientific community would provide policymakers with up-to-date information on the dynamics of the Earth's natural and socio-economic systems, and identify where uncertainties exist or where the evidence base is inadequate, in order that informed decisions are made at both the national and international levels (The Royal Society, 2010: 5).

Science in diplomacy, in other words, is about equipping international decision-makers with the scientific knowledge and understanding required to cope with the increasingly complex S&T-related demands of the 21st century. More and more foreign policy decisions are drawing on information that science and the scientific community provides. In looking at current challenges, such as those related to global health, climate change, weapons proliferation and economic growth and innovation, it must be acknowledged that science, technology and knowledge have potentially a central role to play in providing possible solutions. None of these issues can be fully addressed without: (1) understanding the science driving the challenge; (2) developing the technical institutions to disseminate information and knowledge about the challenge; and (3) engaging with technical experts. As such, decision makers need access to both highly qualified people and timely and relevant information.

The establishment of the Intergovernmental Panel on Climate Change (IPCC) is probably one of the better known examples of policy-related scientific advice, and is a contemporary illustration of science in diplomacy. Mechanisms have been established to aid the flow of information regarding climate change and its potential consequences — from the environmental to the economic — from global scientific institutions and research bodies to the policy making community. While the IPCC does not carry out original research, it reviews and produces periodic assessments of recent scientific, technical and socio-economic research from around the world, and differing viewpoints from within the scientific community are reflected in its reports.

These reports have had far reaching effects in the realm of international relations and on the activities of scientific institutions. Affiliated bodies such as the World Meteorological Organization and the United Nations Environment Programme (which together established the IPPC in 1988), also influence international relations. Scientific knowledge informed the 1992 United Nations Framework Convention on Climate Change and the 1997 Kyoto Protocol, both of which stipulated binding obligations for states to reduce carbon emissions. The notion of 'carbon credits', and the process of states offsetting the limitations of their pollution by purchasing carbon credits from other states, highlight the impact of science in the policy sphere. In December 2007, the IPCC was awarded the Nobel Peace Prize (jointly with former US Vice-President Al Gore) 'for their efforts to build up and disseminate greater knowledge about man-made climate change, and to lay the foundations for the measures that are needed to counteract such change' (Nobel Media AB, n.d.).

National and international academies, learned societies and national scientific advisory bodies are also important sources of independent scientific advice for foreign policy makers. For example, the InterAcademy Panel on International Issues (IAP) — which represents more than 100 national science academies around the word — published statements on ocean acidification and deforestation as part of the UN climate change negotiations in 2009 (The Royal Society, 2010: 6). A decade earlier, a report by the US National Academy of Scientists concluding that the majority of US foreign policy objectives had science, technology and/or health implications led to the appointment of a science advisor to the Secretary of State and a more than 15-fold increase in the number of scientists with PhDs receiving fellowships to work in the State Department or USAID. Similarly, the Obama administration has recruited several Nobel laureates to fill key executive branch positions, including Secretary of Energy Steven Chu (National Academy of Sciences, National Academy of Engineering, and Institute of Medicine, 2011: 8). Such developments have helped the US foreign policy community build stronger links to the US scientific community and increased America's overall scientific capacity to deal effectively with the many technical issues that arise in contemporary US foreign policy. This pattern is not replicated to the same degree everywhere in the world, but there are signs of increased scientific input in policy-making in general and in diplomacy in particular.

Finally, building stronger inter-agency collaborations, so that foreign policy makers have easier access to the pool of technical knowledge and communities available in government ministries or departments, and fostering stronger scientific civil societies that have the ability to formally or informally advise international policy leaders, will continue to be objectives of key importance in the future. Ultimately, the ability for science and scientists to equip decision makers with necessary technical and scientific information and also the willingness of decision-makers to recognise the need for such information will help determine the effectiveness of international responses to some of the world's most pressing challenges.

Science for diplomacy

> Science diplomacy and science and technology cooperation ... is one of our most effective ways of influencing and assisting other nations and creating real bridges between the United States and counterparts.[11]

Unlike the categories above, science *for* diplomacy is the use of science to help build and improve international relations, especially where there may be

[11]Hillary Clinton, US Secretary of State (2009) as quoted in (National Academy of Sciences, National Academy of Engineering, and Institute of Medicine, 2011: 8).

strain or tension in the official relationship. Science for diplomacy primarily draws on the 'soft power' of science: its attractiveness and influence both as a national asset, and as a universal activity that transcends national or partisan interests.

By enabling countries to exercise and express 'soft power' in new and highly effective ways, and fostering the development of trust and agreement between often-adversarial nations, the science for diplomacy dimension is increasingly acknowledged to be of real potential significance. In describing the importance of his country's research and discoveries in its broader global strategies, Professor Peter Gluckman, Chief Science Advisor to New Zealand's Prime Minister, said: 'As a small nation we must compete hard to maintain our relevance in a world where we can easily be forgotten. We have to demonstrate that small countries can indeed, make a difference' (Gluckman, 2011), in his address at the 1st Annual Meeting of the New Zealand Greenhouse Gas Research Centre. A country's attempts to project influence and importance on a global scale through its national scientific community provides a fascinating snapshot of science for diplomacy in action. Other nations are also picking up on this potential power. For example, through its 'science without borders' initiative (now known as The Brazil Scientific Mobility Program), Brazil is not only training future scientists internationally but is critically using science as a way to reach out to key strategic allies and important economic partners.[12] Other countries, such as China, are using large investments in science and infrastructure both to build their national science systems and to reach out to, and attract, top talent from around the world to their shores (Marcelli, 2013).

Perhaps the real promise of science for diplomacy, however, lies in its ability to develop stronger links between countries in which the political

[12]The programme is part of the Brazilian government's broader effort to grant 100,000 scholarships to the country's top undergraduate and postgraduate students in the Science, Technology, Engineering and Mathematics (STEM) fields to enable them to study abroad at the world's best universities. Jointly funded by CAPES (an organisation within the Brazilian Ministry of Education) and CNPq (an organisation within the Brazilian Ministry of Science and Technology), the initiative aims to promote scientific research, increase international cooperation within science and technology, and to engage students in a global dialogue. For more information see IIE (n.d.).

environment is tense and official relationships are strained or limited.[13] The emergence of an era of science diplomacy — in which non-governmental scientists and academics play a key role in diplomacy and international policy — has already provided US scientists with access to potentially influential communities in countries such as Cuba, Burma, Iran and North Korea, despite recurring political crises and the absence of formal government-to-government relations. In particular, initiatives undertaken by the US National Academy of Sciences (NAS) in Iran, in areas such as earthquake science and food-borne diseases, have provided one of the few enduring links between the two countries over a decade marked by particular distrust and tension.[14] Similarly, university partnerships, such as the nearly ten year-long collaboration between Syracuse University and Kim Chaek University of Technology in North Korea on standards-based information technology (Thorson *et al.*, 2008), have enabled people-to-people contacts to persist despite the near total lack of sustained connections at the official diplomatic level between their respective nations.

Like other dimensions of science diplomacy, science for diplomacy comes in many forms. These include, but are not limited to:

- *Science cooperation agreements.* Agreements, such as that signed between Libya and the US in 2004 after the former consented to relinquish its WMDs, are often used to symbolise thawing political relations.
- *Creation of new institutions.* International academies and institutions, such as the European Organisation for Nuclear Research (CERN), can be specifically created in order to reflect and promote the goals of science for diplomacy.
- *Educational scholarships.* Educational scholarships and exchanges act as a mechanism for network-building, and encourage global partnerships. The Newton International Fellowships, run jointly by the Royal Society,

[13]This characteristic helps to explain the current focus within US foreign policy on expanding science diplomacy with the Arab and Islamic worlds, and it aptly illustrates the use of *science for diplomacy* (Lord and Turekin, 2007). For an extraordinary, but all too rare, multilateral example is the SESAME Synchrotron project in Jordan, see Smith (2012).

[14]For more information on the use of science diplomacy to foster US-Iranian engagement see Jillson (2013) and Albro (2014).

the Royal Academy of Engineering and the British Academy, are a case in point.
- *'Track II' diplomacy.* In contrast to 'track I', or official, diplomacy, 'track II' diplomacy directly involves those — such as scientists and other academics — working outside of the official negotiation process.
- *Science festivals and exhibitions.* These events often constitute an effective platform from which to emphasis the universality and impartiality of science, and to highlight common interests. Countries such as China, India and Iran are particularly proud of their historical contributions to scientific advancement, and are keen to share and celebrate this with the world.

Exploring the Significance of Science Diplomacy

This book seeks to do more than acknowledge the emergence of Science Diplomacy in the international arena. It also attempts to examine the significance of this development and assess whether the advent of Science Diplomacy represents a major break from the past. The structure of this book reflects this central concern.

The first four chapters in this volume focus on the theme of Diplomacy for Science. Using President Barack Obama's commitment of June 2009 to expand science and technology engagement with the Muslim world as a benchmark, Cathleen A. Campbell outlines specific initiatives taken to advance US science diplomacy in the Arab world since 2009 and then pinpoints some key lessons of this experience. Sarah Macindoe assesses current international efforts to manage plant genetic resources for food and agriculture and whether New Zealand can harness science diplomacy to make a positive impact here. For Gary Wilson, Antarctica has a critical role in world's ocean and atmospheric system and it is now imperative for the model of international co-operation, based on links between science and diplomacy on the frozen continent, to be extended to counter the threat of global warming. In addition, Maria Pozza examines the Square Kilometre Array (SKA) radio telescope project as a case that is not only deepening scientific links between South Africa and Australia (and to a lesser degree, New Zealand) but is also expanding diplomatic links between a developing and developed state.

Four subsequent chapters deal with various aspects of Science in Diplomacy or how scientific advice interacts with foreign policy goals. Manjana Milkoreit explores the fascinating question of how scientific information is received and used in the minds of diplomats by probing the belief systems of diplomatic participants in the United Nations Framework Convention on Climate Change (UNFCCC) negotiations. Drawing on his own high-level professional experience, Sefton Darby looks at the international hydrocarbon and minerals extraction environment in two countries — Chad and Azerbaijan — and, in particular, considers the relationship between the 'resource curse' and science diplomacy. Joan Leach outlines the problems and possibilities for science communication in international diplomacy. Science communication is considered as a form of 'soft power' in the three related areas of diplomacy for science, science in diplomacy, and science for diplomacy that characterise science diplomacy. In contrast, Daryl Copeland focuses on the role of science and technology in today's world and looks at the 2010–2011 WikiLeaks 'Cablegate' affair as a case study of the impact of digital communications technology on contemporary diplomacy *(technology in diplomacy).*

The final four chapters provide insights into the possibilities and challenges of Science for Diplomacy. In a chapter concerning the association between science cooperation and international security, Jeffrey Boutwell explores the impact of the information and communications revolution on three 21st century security issues — missile defence, militarization of outer space, and the geopolitics of the Artic in the era of climate change. Meanwhile, Edison T. Liu considers global health research as a specific form of science diplomacy and drawing on three examples — epidemic research, clinical cancer research and population genetics research — he maintains that this form of collaboration delivered substantial and diplomatic benefits. Stephen Goldson and Peter Gluckman consider how a small state like New Zealand, a predominantly food exporting nation, strategically uses science to maximise diplomatic impact in seemingly diverse areas such as biosecurity and pastoral gas greenhouse emissions. Furthermore, Atsushi Sunami, Tomoko Hamachi, and Shigeru Kitaba analyse a growing recognition in Japan that science and technology diplomacy has a big role to play if Tokyo is to remain one of the 'critical points' in an expanding global science resource network.

References

Albro, R. (2014) *Science diplomacy can pave the way between Iran and US*, 25 January, [Online], Available: http://irdiplomacy.ir/en/page/1927842/Science+Diplomacy+can+pave+the+way+between+Iran+and+US.html [24 February 2014].

Barder, O. and Krylova, P. (2013) *Commitment to Development Index 2013*, 18 November, [Online], Available: http://www.cgdev.org/sites/default/files/CDI2013/cdi-brief-2013.html [23 February 2014].

Clark, I. (1997) *Globalization and Fragmentation*, Oxford: Oxford University Press.

Gluckman, P. (2011) *New Zealand Science and our International Connections: Science 'globalisation' is the future*, 22 February, [Online], Available: http://www.mahurangi.org.nz/Climate/Science-Globalisation-is-the-Future.php.

Harding, T.K., Khanna, M.J. and Orbach, R.L. (2012) 'International fusion energy and cooperation: ITER as a case study in science and diplomacy', *Science and Diplomacy*, 9 March, [Online], Available: http://www.sciencediplomacy.org/article/2012/international-fusion-energy-cooperation [14 February 2014].

Held, D. and Anthony, M. (1999) *Global Transformations*, Cambridge: Cambridge University Press.

Held, D., McGrew, A., Goldblatt, D. and Perraton, J. (1999) *Global Transformations: Politics, economics and culture*, Standford: Standford University Press.

International Institude of Education (IIE) (n.d.) *Brazil Scientific Mobility Program*, [Online], Available: http://www.iie.org/Programs/Brazil-Scientific-Mobility/About [23 February 2014].

Jillson, I.A. (2013) 'The United States and Iran: Gaining and sharing scientific knowledge through collaboration', *Science and Diplomacy*, 18 March, [Online], Available: http://www.sciencediplomacy.org/article/2013/united-states-and-iran [24 February 2014].

Keohane, R. and Nye, J. (2000) 'Globalization: What's New? What's Not? (And So What?)', *Foreign Policy*, Spring 2000(118): 104.

Lord, K.M. and Turekin, V.C. (2007) 'Time for a New Era of Science Diplomacy', *Science*, 315(5813): 769–770. DOI: 10.1126/science.1139880.

Marcelli, A. (2013) 'The large research infrastructures of the People's Republic of China: An investment for science and technology', *Physics Status Solidi B (Basic Solid State Physics)*. DOI: 10.1002/pssb.201350119.

National Academy of Sciences, National Academy of Engineering, and Institute of Medicine (2011) *Examining Core Elements of International Research Collaboration: Summary of a Workshop*, Washington, DC: The National Academies Press.

National Research Council (2002) *Knowledge and Diplomacy: Science advice in the United Nations system*, Washington, DC: The National Academies Press, Available: http://www.nap.edu/catalog.php?record_id=10577.

Nobel Media AB (n.d.) *The Nobel Peace Prize 2007*, [Online], Available: http://www.nobelprize.org/nobel_prizes/peace/laureates/2007/ [14 February 2014].

Nye, J. (2004) *Soft Power: The means to success in world politics*, New York: Public Affairs.

Perkins, N.I. (2012) *Is the World Ready for Science Diplomacy?*, 12 October, [Online], Available: http://www.scidev.net/global/science-diplomacy/editorials/is-the-world-ready-for-scientific-diplomacy--1.html [23 February 2013].

Prolavorio, C. (2013) *Marking over 50 Years of Franco-German Cooperation*, 2 April, [Online], Available: http://home.web.cern.ch/students-educators/updates/2013/04/marking-over-50-years-franco-german-cooperation [24 February 2014].

Regan, R. (1985) *Address to the Nation on the Upcoming Soviet-United States Summit in Geneva*, 14 November, [Online], Available: http://www.reagan.utexas.edu/archives/speeches/1985/111485d.htm [23 February 2014].

Reischauer, E.O. (1960) *The Broken Dialogue with Japan*, October, [Online], Available: http://www.foreignaffairs.com/articles/71579/edwin-o-reischauer/the-broken-dialogue-with-japan.

Scholte, J. (2001) 'The globalization of world politics', in Baylis, J. and Smith, S. (eds.) *The Globalization of World Politics: An introduction to international relations*, 2nd edition, Oxford: Oxford University Press, pp. 13–32.

Science and Technology Facilities Council (STFC) (n.d.) *Large Hadron Collider (LHC)*, [Online], Available: http://www.stfc.ac.uk/646.aspx [23 February 2014].

Smith, C.L. (2012) *Synchrotron light and the Middle East: Bringing the region's scientific communities together through SESAME*, 16 November, [Online], Available: http://www.sciencediplomacy.org/perspective/2012/synchrotron-light-and-middle-east.

The Royal Society (2010) *New Frontiers in Science Diplomacy: Navigating the changing balance of power*, January, [Online], Available: https://royalsociety.org/~/media/Royal_Society_Content/policy/publications/2010/4294969468.pdf.

Thorson, S., Harblin, T. and Carriere, F.F. (2008) 'US-North Korea trust building through academic science cooperation', *Journal of the World Universities Forum*, 1(3): 57–64.

Tumushabe, G.W. and Omar-Mugabe, J. (2012) *Governance of Science, Technology and Innovation in the East African Community: Inaugural Biennial Report 2012*, [Online], Available: http://www.ist-africa.org/home/files/GovernanceofSTIEast AfricaCommunity_2012.pdf.

Turekin, V.C. and Neureiter, N.P. (2012) 'Science and Diplomacy: The past as prologue', *Science and Diplomacy*, 9 March, [Online], Available: http://www.sciencediplomacy.org/editorial/2012/science-and-diplomacy.

TVNZ (2011) *Scientists top 'most trusted' list*, 20 June, [Online], Available: http://tvnz.co.nz/national-news/scientists-top-most-trusted-list-4247442.

Wade, N. (1974) 'Kissinger on Science: Making the link with diplomacy', *Science*, 184(4138): 780–781. DOI: 10.1126/science.184.4138.780.

Wagner, C. (2008) *The New Invisible College: Science for Development*, Washington, DC: Brookings Institution Press.

Weizmann Institute of Science (2013) *Weizmann Institute and Max Planck Society Establish a Joint Centre for Archaeology and Anthropology*, 11 January, [Online], Available: http://wis-wander.weizmann.ac.il/weizmann-institute-and-max-planck-society-establish-a-joint-center-for-archaeology-and-anthropology#.Uv1Wi_mSyzs [14 February 2014].

Part I

Science in Diplomacy: Facilitating International Science Co-operation

CHAPTER 2

US Science Diplomacy with Arab Countries

Cathleen A. Campbell

In Cairo, Egypt in June 2009, US President Barack Obama delivered a pivotal speech in which he outlined a 'new beginning' in US engagement with the Muslim world. He announced a wide range of initiatives that included a commitment to expand science and technology engagement. President Obama specifically called for the creation of centres of excellence and for increased cooperation in health. He urged partnerships and investment to assist entrepreneurs to create new businesses and jobs. He also announced the appointment of US 'science envoys' to collaborate with Muslim countries on programs related to energy, agriculture, information technology and water (Obama, 2009).

At first glance, President Obama's call to action appears similar to previous US presidents' calls for increased international cooperation in science and technology. Presidents and prime ministers often urge cooperation in science during overseas speeches. International summits and ministerial meetings often result in new agreements in science and technology. In this particular instance, however, the push for more science and technology engagement with Muslim countries was indeed tied to the desire to improve relations with those countries. A report issued in February 2009 by WorldPublicOpinion.org, based on the results of in-depth surveys in three countries supplemented by worldwide polling, showed strong negative views of the US government within Muslim countries. The study found that then '...the US is widely seen as hypocritically failing to abide by international law, not living up to the role it should play in world affairs, disrespectful of the Muslim people, and using its power in a coercive and unfair fashion'

(Kull *et al.,* 2009). Polling done by the Pew Centre consistently showed strong admiration for US science and technology, even in countries where the overall views of the United States are low. For example, in a 2007 survey, 83 percent of Malaysians, 68 percent of Jordanians, and 67 percent of Palestinians expressed a positive view of US science and technology (Kohut and Richard, 2009). Expansion of cooperation in an area where there is respect for US progress and a desire to learn from the US experience provided an opportunity — at least in principle — to help improve the perception of the US in Muslim majority countries. Scientists and policy leaders in the region also recognize this potential. In October 2011, the Islamic World Academy of Sciences (IAS) convened its 18th international conference in Doha, Qatar under the theme: *The Islamic World and the West: Rebuilding bridges through science and technology.* One of the three conference objectives focused on hearing from the over 200 participants about '...ways to bridge the divide between the Islamic World and West, particularly through science and scientific and technological collaborations' (IASWorld, 2012: 3).

President Obama's June 2009 speech raised expectations about the potential of science engagement to improve relations. As in many complex situations, the reality is much harder to achieve. The difficulty of turning President Obama's grand vision into actual programs, the challenge of mobilizing resources in times of financial constraints, and the complexities inherent in trying to implement programs in countries experiencing ongoing political and economic turmoil, have resulted in slower implementation than many anticipated.

In his Cairo speech, and in follow-up actions, the Obama Administration has focused on improving relations with Muslims worldwide, including those living in Asia, the Middle East, Africa and Europe. The Muslim world encompasses more than fifty countries that vary greatly in population, economic output, natural resources, education and scientific output. This chapter will focus on a subset of countries that due to longstanding political and economic relationships, as well as the legacy of 9/11, have special significance to the United States. This chapter will address US science diplomacy with 20 Arab countries[1] in the Middle East, Africa and the Gulf region. Other

[1] Algeria, Bahrain, Djibouti, Egypt, Iraq, Jordan, Kuwait, Lebanon, Libya, Mauritania, Morocco, Oman, Palestine, Qatar, Saudi Arabia, Sudan, Syria, Tunisia, UAE, and Yemen.

Muslim-majority countries, such as Indonesia, Pakistan, Iran, and Malaysia will not be discussed in depth. However, they may be referenced for comparison or be included in selected data sets.

Defining Science Diplomacy

Recognition of science diplomacy as an important component of international relations is growing. Policymakers around the world are referencing science diplomacy in speeches; experts are convening conference panels or workshops on the topic. Press interest is growing; and there is a new journal devoted to the topic — *Science and Diplomacy*, published by the Centre for Science Diplomacy of the American Association for the Advancement of Science (AAAS). This recent surge in writing and analysis is welcome, particularly if it helps to shape a common understanding of science diplomacy, enable those involved in science diplomacy to document and share lessons learned from their experiences, and lay groundwork for establishing a common framework for assessing science diplomacy's impact.

While progress is being made in documenting and sharing lessons learned, less progress is being made in reaching a common definition of science diplomacy and developing a common framework for assessing its impact. To some, science diplomacy refers to any type of international science cooperation, including well-established government-to-government cooperation between allies or any multilateral science cooperation such as that taking place at the Large Hadron Collider in Europe where nearly 8,000 scientists and engineers representing 60 countries are working. Indeed, the Large Hadron Collider project — particularly the intergovernmental involvement in negotiating the large-scale collaboration, can be viewed as an example of 'Diplomacy for Science', which AAAS and the UK Royal Society define as the process of facilitating international science cooperation through diplomacy'. Two other dimensions of science diplomacy, as defined by AAAS and the Royal Society, include 'Science in Diplomacy', which is 'informing foreign policy with science advice', and 'Science for Diplomacy', which is 'the process of using science cooperation to improve international relations' (The Royal Society, 2010).

Others have adopted this broad-brush approach to science diplomacy. In an August 2013 speech, the Science and Technology Advisor to the

US Secretary of State defined science diplomacy as: '(1) science and technology aiding diplomacy (for the many diplomatic issues where scientific and technological information is critically important and even for those cases where science and technology engagement can open doors for dialogue on other issues); (2) diplomacy advancing science and technology (such as by negotiating multinational arrangements for building large facilities and gaining access for research in unique locations) and (3) science and technology helping to solve national, regional and global problems (such as by creating new options and paths for making progress on the "wicked problems" too difficult for politicians to resolve alone)' (Colglazier, 2013).

The AAAS Centre for Science Diplomacy, established in 2008, takes a more nuanced approach. Its website clearly highlights the relationship-building benefits of science diplomacy. Further, it notes the Centre's interest in science diplomacy serving '...as a catalyst between societies where official relationship might be limited and to strengthen civil society interactions through partnerships in science and technology' (AAAS, 2014). It is this potential to improve relations through science engagement that gives special meaning to science diplomacy and sets it apart from more traditional international science and technology cooperation.

However, this approach to science diplomacy makes it difficult to monitor and evaluate its outcomes and impacts. An important value of science diplomacy is building trust and confidence between countries, particularly those in conflict or with strained relations. But trust is difficult to measure and the impact that those trusted relations have on improving relations between countries more broadly is even harder to measure. Advocates of science diplomacy often cite the dialogue between US and Soviet scientists during the Cold War as an example of science diplomacy. Scientists and engineers involved in that dialogue — both American and Russian — believe that the relationships and understanding developed during those years enabled the two communities to successfully collaborate on key science and security challenges after the collapse of the Soviet Union. Equally important was the fact that some of these scientists (particularly in Russia and other Eurasian countries) had reached high-level government positions and the experience they had working with US counterparts paved the way for compromise and collaboration on key issues. But this process and these relationships evolved over many years, suggesting that any potential impacts of science diplomacy efforts with Arab countries will take years, if not decades, to realize.

Potential for Science Diplomacy with Arab Countries

Although the US has cooperated with many Arab countries for years, there is considerable room to expand existing collaboration and to use science diplomacy to engage countries with which connections are weak and to work collaboratively to solve common challenges. Arab countries share strong cultural and linguistic links but vary greatly economically, politically, socially and, of course, scientifically. While there are some exceptions, Arab countries generally lag behind other countries in terms of science infrastructure, investment, output, and collaboration with other countries. Working in partnership with these countries to help build indigenous science and engineering capacity and increase global integration represents a significant science diplomacy opportunity.

Arab countries comprise a very small share of global expenditures on research and development (R&D). At the individual country level, R&D intensity (the gross domestic expenditure on R&D as a percentage of GDP) generally falls far below the world average (2.2 percent). Figure 1 shows that only Tunisia exceeds the world average and only Tunisia and Morocco have met the 1 percent target established by the Organization of Islamic

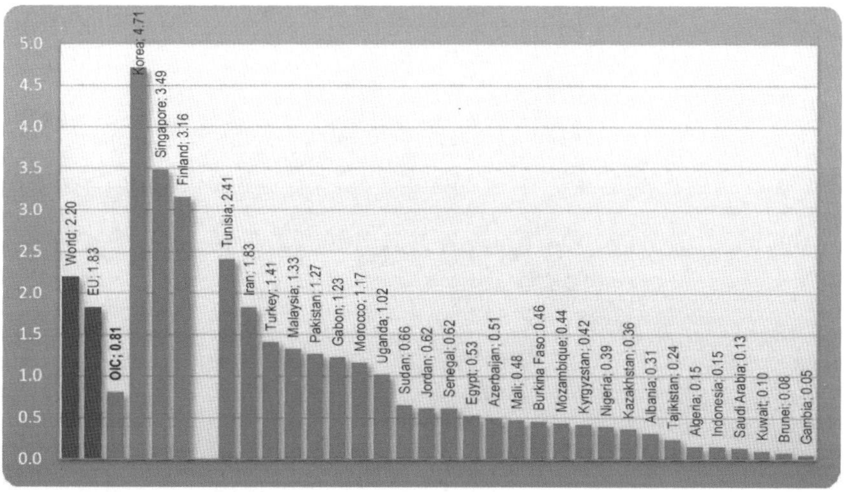

Figure 1. R&D Intensity (in percentage).

Note: Data for the most recent year available between 1999 and 2010. Image taken from OIC (2011: 4–6).

Cooperation (OIC) for its members (all of the 20 Arab countries featured in this chapter are members of the OIC).

Similarly, R&D expenditures per capita lag other regions. Among Arab nations, Tunisia ranks highest, at US$100.5million. The second highest Arab country — Kuwait — spends considerably less per capita (US$31.5million). All other Arab countries fall below the OIC average of US$27.7milliom, and far below the world average of US$219million and the EU average of US$601million (OIC, 2011: 6).

The number of scientific researchers as a percentage of the total population is also quite low. For all OIC countries the average number is 451 scientists per million people, which compares poorly against the global average of 1,507 per million inhabitants. Figure 2 shows significant disparity among Arab countries. According to the UNESCO Science Report 2010, which used survey data to estimate the number of researchers based on full-time equivalents, Jordan and Tunisia have the highest number of researchers among the Arab countries (Badran and Zou'bi, 2010: 261).

One of the key indicators used to assess research productivity is the number of scientific articles published in indexed journals. Scientific publications per million of the population in the Arab world reached 13,754 in 2008, led by Kuwait, Tunisia, Jordan, Qatar and UAE. This almost doubles the number of publications (7,446) in 2000, but is well below that of global

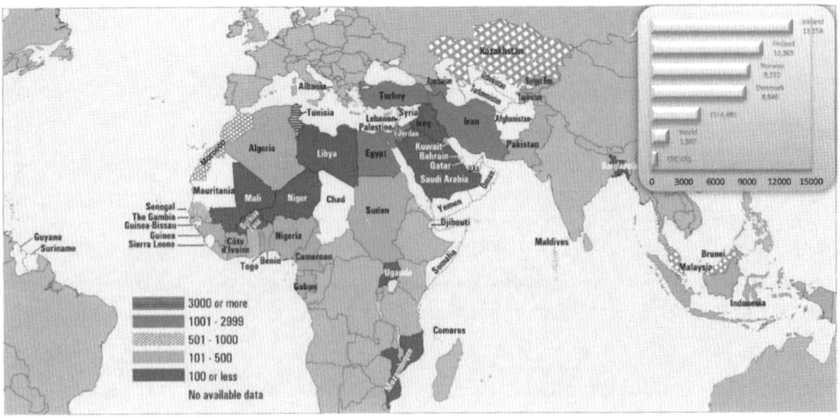

Figure 2. Researchers per million people.

Note: Headcount data for the most recent year available. Image taken from (OIC, 2011: 2).

leaders, including the US (323,000), China (137,000) and Germany (85,000) (Organisation of Islamic Cooperation (OIC, 2011: 12).

Science is increasingly a global enterprise and global integration is a necessary component of strong national science systems. Success in publishing research results in international, peer-reviewed journals is one indication of global integration. According to the US National Science Foundation, the share of the world's science and engineering articles with international co-authors grew from eight percent in 1988 to 23 percent in 2009. For some countries or regions, the percentages were much higher. For example, the share for the US grew from about 11 percent in 1989 to 31 percent in 2009; EU grew from about 18 percent to 42 percent in 2009; and China grew from about 23 percent in 1989 to 26 percent (National Science Board, 2012).

The UNESCO Science Report of 2010 shows an increase in co-publications involving scientists from the region. Citing data from Thomson Reuters Inc., the report notes that between 2000 and 2008 there was a steady increase in the number of Arab scientists collaborating with the diaspora. For example, approximately one third of the 3,963 articles published by Egyptian scientists in 2008 were co-authored by scientists outside Egypt. Similar trends were seen in Tunisia, Saudi Arabia, Algeria, UAE, Jordan and Lebanon (UNESCO, 2010: 264–267).

The above data demonstrate that, despite some progress in the last decade, Arab countries lag significantly in their production and use of indigenous science and technology knowledge. Across the board, Arab countries that are interested in creating a knowledge-based economy need to invest more in education, research and entrepreneurship, and in creating the necessary infrastructure and policy frameworks to support science, technology and innovation. These are areas where the United States, with its unique experience in fostering science and technology based economic growth, can utilize science diplomacy to improve relations with the Arab world. However, US science diplomacy efforts must be carefully aligned with the interests, capabilities, and absorptive capacity of the partner country. The increased attention that many Arab countries are devoting to science, technology and innovation suggest that the potential alignment exists.

Scientific and political leaders in many Arab countries are working to improve indigenous science capabilities and also to spur collaboration and integration with the global community. For example, Saudi Arabia is investing billions to create a modern, global research university — the King Abdullah University for Science and Technology (KAUST) — that is attracting top-notch faculty and postgraduate students from around the world. Opened in 2009, KAUST aims to reach 250 faculty and 2,000 students by 2020. The Saudi government continues to support other initiatives to strengthen and expand science and technology, including the King Abdul Aziz city for Science and Technology and other initiatives to promote technology transfer (The Royal Society, 2010: 15). Qatar aims to spend 2.8 percent of GDP on research by 2015, building on its recent educational initiatives, such as the creation of Education City, and support for R&D through the Qatar National Research Fund (all under the Qatar Foundation). The United Arab Emirates is creating 'the world's first fully sustainable city and innovation hub' through its Masdar Initiative, which is being developed in partnership with the Massachusetts Institute of Technology (MIT) (The Royal Society, 2011: 22).

In addition to numerous national initiatives to boost science, technology and innovation, there have been a number of multinational efforts within the Arab world and within the OIC. To date, these have been less successful than national efforts. In 2010 the heads of state attending the Arab Summit adopted a resolution mandating the development of an S&T strategy for the entire Arab region. In 2010, the OIC adopted 'Vision 1441' which is a long-term blueprint for the contribution of science, technology and innovation to economic and social development across the OIC. The plan calls for significant increases in the number of students attending university, the percentage of gross domestic product invested in research and development, and innovation activity. Vision 1441 — referring to a target year in the Islamic calendar that coincides with the year 2020 in the Gregorian calendar — challenged OIC countries to produce 14 percent of the world's scientific output; to produce 1,441 researchers, scientists, or engineers per million of their population; and to devote 1.4 percent of their gross domestic product to research and development (Osama, 2011).

Another reason for optimism in the potential of science diplomacy is the fact that Arab countries, like countries everywhere, face a number of long-term challenges that require science and technology to solve. The UNESCO 2010 World Science Report listed energy and water security as top R&D priorities in Arab countries, and underscored the need for Arab countries to address the World Summit on Sustainable Development's five priority areas of water, energy, health, agriculture, and biodiversity. In many cases, specific problems are cross-border in nation and require cross-border collaboration to solve. Progress in these areas could have long-lasting impact on the economy, the health and well-being of its citizens, and the reduction of conflict within the region.

New Beginning

Since 2009, the US government has launched or catalysed numerous bilateral, regional and multi-national initiatives to expand science and technology engagement with the Muslim world. Many build on existing science cooperation implemented by US government federal science agencies. However, the list also includes several new initiatives, the status of which is described below (OSTP, 2010). Combined, this list of initiatives demonstrates a concerted effort to follow through on the commitments of the Cairo speech. However, the scope and impact are considerably less than what is needed to utilize science diplomacy to positively affect relations with these countries.

Science envoys

Beginning in 2009, the US government has named nine prominent US scientists to serve as Science Envoys. In coordination with the Department of State, these envoys have travelled to Egypt, Morocco, Tunisia, South Africa, Ethiopia, Indonesia, Kazakhstan, Uzbekistan, Bangladesh and other countries to explore ways to strengthen partnerships and solve common science and engineering challenges. Upon their return, the science envoys offer recommendations for collaboration and, very importantly, serve as advocates for specific follow-up activities.

In 2010, the first three Science Envoys appeared before the US President's Council of Advisors on Science and Technology (PCAST) to discuss science diplomacy. All three pointed to early progress achieving the vision articled in President Obama's Cairo speech. Dr Bruce Alberts, Science Envoy to Indonesia, outlined four challenges: (1) defining a role for the science envoy that demonstrates the effectiveness of science diplomacy; (2) creating a science envoy 'toolkit'; (3) helping the US government create a structure that maximizes the benefit of the science envoy; and (4) advocating the value of science envoys to all major nations. Dr Elias Zerhouni, the Science Envoy to Algeria, stressed the potential and desire of many countries to learn from the US experience in science and technology. He also argued that visa and security restrictions imposed after 9/11 had prompted some Arab countries to turn toward Europe and Asia rather than the United States (Boisseau, 2010).

The observations by the science envoys underscore the opportunity and challenges this particular science diplomacy initiative presents. On the one hand, the appointment of such highly-regarded and accomplished scientists with direct knowledge of the countries to which they were assigned gave tremendous credibility to the program. But it also raises expectations that are difficult to meet. The science envoy program has no funding associated with it, which means that the envoys have to focus on convincing US government agencies to allocate resources for follow-up activities. Finally, while from a foreign policy perspective the expansion of the science envoy program to other countries has value, it also dilutes the potential impact of this initiative to improve relations with the Arab world. The State Department's announcement of the third class of science envoys in November 2012 references science diplomacy globally and does not mention any specific country or region where the envoys would focus (US State Department, 2012).

Centres of excellence

Since 2009, the US Agency for International Development and the State Department began the creation of two centres of excellence. One centre, in the Middle East, focuses on water and a second centre, in Asia, focuses on climate change.

Entrepreneurship summit

In April 2010, President Obama convened the first global Entrepreneurship Summit, which brought together successful business and social entrepreneurs, venture capitalists, bankers and business experts to discuss ideas and share experiences with a view to creating networks to support entrepreneurs in Muslim communities. The Governments of Turkey, UAE and Malaysia convened the subsequent annual Entrepreneurship Summits in 2011, 2012, and 2013, respectively.

Frontiers of science program

The National Academy of Sciences and the National Academy of Engineering have introduced an 'Arab-American Frontiers of Science, Engineering, and Medicine program'. This program convenes outstanding young scientists, engineers, and medical professionals from the United States and the 22 countries of the Arab League. The stated goal is to increase scientific exchange and dialogue among young researchers in Arab countries and the United States, and to facilitate research collaboration within and beyond the region. The first symposium, done in partnership with the Kuwait Institute for Scientific Research, was held in October 2011. The second symposium is planned for December 2014 in partnership with The Research Council (TRC) of Oman (US National Academy of Sciences, 2014).

The Frontiers of Science program demonstrates the role that non-governmental organizations can play in improving relations through science. Science diplomacy is well-suited to non-governmental actors who often are better able than government agencies to engage with counterparts in countries with whom official relations are strained. Indeed, the early US examples of science diplomacy with the Union of Soviet Socialist Republics and with the Peoples' Republic of China demonstrate the key role that individuals and organizations outside of government played. They established dialogue and early exchanges that laid a foundation for the formal government-to-government relationships that followed (Campbell, 2012). More recent examples include the work of the US-DPRK Consortium for Science Engagement, and the science diplomacy initiatives that the American Association for the Advance of Science and others are pursuing with Myanmar (Daniels, Thet Khin and Agre, 2012). Additionally, the US National Academies of Sciences has implemented an impressive 12-year program of engagement between US and Iranian scientists and engineers. During the first decade of this program, workshops involving more than 500 scientists and engineers from over 80 institutions in the United States and Iran were held in eight priority areas: food-borne diseases; effective use of water resources; earthquake science and engineering; science, ethics, and appropriate uses of technology; science and society; preventing and responding to crises; ecology and energy; and higher education and research challenges (National Research Council, 2010).

Science diplomacy is also a key focus of my organization, CRDF Global. Established in 1995 as an independent, non-governmental organization, CRDF Global's mission is to promote peace and prosperity through international science cooperation. CRDF Global supports science for discovery (through collaborative research and capacity building); science for economic growth (through innovation and entrepreneurship programs); and science for safety and security (through specialized training and research).

CRDF Global was established as a result of legislation that the US Congress passed in 1992 to deal with the aftermath of the dramatic changes in Eurasia. Congress recognized that a nongovernmental organization could operate more flexibly and quickly than US government agencies in getting programs established. A nongovernmental organization could work with a wide range of partners in other countries including government agencies, universities, businesses and other non-governmental organizations. Plus, a nongovernmental agency could more easily leverage resources within the US and abroad to support priority programs.

Today CRDF Global is active in more than 40 countries in Eurasia, the Middle East, North Africa and Asia. Many of these countries are experience changes that are as cataclysmic as the fall of the Berlin wall in 1989 and the changes that resulted in Eurasia and Eastern Europe. If that earlier period is any guide, now is absolutely the right time to be thinking about science diplomacy and the role that it can play in whatever transitions that occur in these regions.

Similar to other nongovernmental organizations, CRDF Global has significantly extended its outreach to scientists and engineers in the Muslim world, including Arab nations. Three innovative programs demonstrate the potential of science diplomacy.

Iraq science engagement

In 2004 CRDF Global invited a small group of Iraqi scientists and engineers to come to the United States for a 'familiarization' trip. This was an important first step in reaching out to a community that had been isolated from the West for years and that was very much in need of support after the end of the Saddam Hussein era. The initial outreach led to the start of lasting relationships; a better understanding of each country's science and engineering systems; the identification of opportunities for collaboration; and subsequent project work. Today, CRDF Global implements, on behalf of the US Department of State, the Iraq Science Fellowship Program (ISFP) which provides opportunities for Iraqi scientists and engineers to research and work alongside colleagues at prestigious host institutions in the US for three-to-six months. Since its inception, CRDF Global has supported the involvement of approximately 50 Iraqi scientists hosted by such notable institutions as the Mayo Clinic; Lawrence Berkeley National Laboratory; Syracuse University; Georgia Institute of Science and Technology, and McKissack & McKissack. The program has achieved great success connecting scientists and engineers to promote peaceful cooperation while achieving real and lasting results. When fellows return to Iraq, their knowledge and new skills are shared with local colleagues and students, amplifying the program's impact. For example, a 2010 engineering fellow, applying concepts learned in the US, successfully proposed a first-of-its-kind renewable energy training centre to an Iraqi Ministry. The proposal was developed in coordination with the fellow's former host and highlights the lasting impact of post-fellowship relationships and the potential of continued collaboration.

Virtual science library

Soon after beginning its engagement with Iraq's science and engineering community, CRDF Global was asked to manage the Iraq Virtual Science Library (IVSL). This innovative project was designed to establish a single portal for Iraq's science and academic communities to access the world's science literature. In addition to providing the servers for the system, developing the portal, and helping to negotiate reduced subscription rates with publishers, CRDF Global conducted numerous trainings for the user community. The impact of the IVSL on Iraq's science community has been remarkable. In 2013, users downloaded an average of 65,000 research articles per month, up from 10,000 per months in 2006. The IVSL now serves more than 80,000 faculty and students from more than 37 universities and research institutions throughout Iraq. In 2010, as planned, CRDF Global handed over management of the IVSL to Iraq. Building on the success of this program, CRDF Global supported a pilot project to create a virtual science library in Afghanistan, the Maghreb countries and Central Asia.

Global innovation through science and technology initiative

One of the most exciting initiatives to engage young scientists and engineers in Islamic countries is the Global Innovation through Science and Technology (GIST) Initiative, implemented in partnership with the US Department of State. Through GIST, CRDF Global seeks to identify, nurture and catalyse young entrepreneurs with a passion for turning innovative ideas into science and technology-based businesses. Working with a network of carefully selected strategic partners overseas, CRDF Global has conducted research and business plan competitions as well as specialized training through start-up boot camps and other workshops. Since launching this program in 2010, CRDF Global has involved thousands of researchers and entrepreneurs in more than forty countries. More importantly, CRDF Global has used this program to link these young entrepreneurs with mentors and business leads both within their own country, region, and globally to help accelerate innovation. With the burgeoning youth populations in Arab countries, this program helps to address an important need — providing economic opportunities for young innovators.

Reality

All of the above initiatives are relatively new and will take time to have impact. As with all science diplomacy initiatives, it will be difficult to demonstrate conclusively that these initiatives, no matter how successful in achieving their own programmatic objectives, help to improve relations between countries in a substantial and sustained way. However, as world leaders continue to face a growing number of challenges that require science and technology-based solutions, the need to engage scientists and engineers will increase.

Meanwhile, there will be important lessons learned as current initiatives unfold. Already we are seeing common challenges in operationalizing science diplomacy. These include the difficulty of obtaining visas for travel between countries; dealing with what typically are complex national regulatory and legal requirements; maintaining direct communications between scientists and engineers involved in projects; and obtaining sustained funding for science diplomacy efforts in situations where there often are prohibitions against using government or even private funds. Developing common approaches or solutions to some of these problems will help to strengthen the case for science diplomacy.

The 4-year history of US efforts to expand science diplomacy with Arab countries suggests several valuable lessons that may prove useful for future initiatives. First, the US initiative was too broad. Science diplomacy is just one of many initiatives launched in the Cairo speech. Even so, science diplomacy was intended to reach over fifty countries with substantially different needs and capabilities. For science diplomacy to be effective, it needs to be focused on a very specific opportunity. Second, achieving success in science diplomacy without financial resources and a commitment to sustain the effort is nearly impossible. The president's speech raised expectations within the US and abroad that the US was prepared to devote substantial financial resources to implement the president's vision. That has not been the case, although the US government deserves credit for mobilizing some resources — including through public-private partnerships — to implement programs. Third, science diplomacy can be effective only if there is political and economic stability. The political and economic challenges of the Arab Spring have required many countries to focus on urgent domestic and security issues and for the US to refocus its foreign policy priorities. In such situations, the opportunities for science diplomacy are more limited.

Nevertheless, scientists and engineers speak a common language and share values that cut across political, economic and social differences. They are able to draw upon shared training and experiences to collaborate on common problems. They can build long-lasting relationships that can serve as a model for others and form a foundation for cooperation in other areas.

From a US perspective, it is clear that the reality of Muslim-world perceptions of the United States will take a very long time to change. The President's Cairo speech raised expectations that many in the region believe have not been met by follow-up action. As a result, based on recent polling, favourable impressions of the US have actually *declined* since 2009 (Pew Research, 2011). So, while the potential for science diplomacy to make a difference still exists, it is clear that there is a lot more to be done. For the US and western countries, the Arab spring offers incredible opportunity to reach out to these countries and to engage scientists in new, creative ways that help to address local needs while building a foundation for long-lasting partnerships.

This is a pivotal time for many countries in the Muslim world. The revolutions that have occurred and may continue to occur in the Middle East and North Africa are changing the landscape, both within the region and internationally, in ways that are reminiscent of the fall of the Berlin Wall more than 20 years ago. Citizens and thought leaders in these countries are wrestling with fundamental questions regarding governance, citizen participation, jobs for the burgeoning youth population, security, etc. We do not know the outcome. We cannot predict what the process will be for getting to whatever outcome emerges. But the interest demonstrated by many science and political leaders in the region to use this unique moment in time to improve governance in their science institutions, to reform university science education, to promote innovation, to raise the quality of science nationally, and to improve the application of science for decision-making and for solving regional and global problems, presents a wonderful opening for science diplomacy.

References

American Association for the Advancement of Science (AAAS) (2014) *About Us — Centre for Science Diplomacy*, [Online], Available: http://www.aaas.org/page/about-0 [27 May 2014].

Badran, A. and Zou'bi, M.R. (2010) 'Arab States', *UNESCO Science Report 2010: The current status of science around the world*, Available: http://www.unesco.org/new/fileadmin/MULTIMEDIA/HQ/SC/pdf/sc_usr10_arab_states_EN.pdf, pp. 250–277.

Boisseau, R. (2010) *FYI: The AIP Bulletin of Science Policy News: PCAST Convenes to Discuss Scientific Diplomacy, Gives Go-Ahead to Health IT Report*, 21 July, [Online], Available: http://www.aip.org/fyi/2010/077.html.

Campbell, C. (2012) 'A Consortium Model for Science Engagement: Lessons from the US-DPRK experience', *Science and Diplomacy*, 28 June, [Online], Available: http://www.sciencediplomacy.org/article/2012/consortium-model-for-science-engagement.

Colglazier, E.W. (2013) *Remarks on Science and Diplomacy in the 21st Century*, 20 August, [Online], Available: http://www.state.gov/e/stas/2013/213741.htm.

Daniels, R., Thet Khin, P. and Agre, P.C. (2012) 'Bringing Health Research to the Renewed US–Myanmar Relationship', *Science and Diplomacy*, 21 June, [Online], Available: http://www.sciencediplomacy.org/editorial/2012/bringing-health-research-renewed-us-myanmar-relationship.

Islamic World Academy of Sciences (IASWorld) (2012) 'Under the patronage of the Prime Minister of Qatar: IAS Convenes 18 Conference in Doha under the title: "The Islamic World and the West: Rebuilding Bridges through Science and Technology"', *Islamic World Academy of Sciences Newsletter*, 26(41), January–March 2012, Available: http://www.iasworld.org/wp-content/uploads/2013/12/Newsletter-41-Final-10-December-2013.pdf.

Kohut, A. and Richard, W. (2009) *Postive Aspects of US Image*, 21 March, [Online], Available: http://hir.harvard.edu/rethinking-finance/positive-aspects-of-us-image.

Kull, S., Ramsey, C., Weber, S., Lewis, E. and Ebrahim, M. (2009) *Public Opinion in the Islamic World on Terrorism, al Qaeda, and US policies*, 25 February, [Online], Available: http://www.start.umd.edu/publication/public-opinion-islamic-world-terrorism-al-qaeda-and-us-policies.

National Research Council (2010) *US–Iran Engagement in Science, Engineering, and Health (2000–2009): Opportunities, constraints, and impacts*, Washington, DC: The National Academies Press.

National Science Board (2012) 'Changing International Research Collaborations', in National Science Foundation *Science and Engineering Indicators 2012*, Available: http://www.nsf.gov/statistics/seind12/pdf/overview.pdf.

Obama, B. (2009) *Reamarks by the President on a New Beginning*, 4 June, [Online], Available: http://www.whitehouse.gov/the-press-office/remarks-president-cairo-university-6-04-09.

Office of Science and Technology Policy (OSTP) (2010) *Fact Sheet: US Government Science and Technology Engagement With the Muslim World: Progress in Realizing the President's Vision of Enhanced Science and Technology (S&T) Partnership in the Muslim World*, [Online], Available: http://www.whitehouse.gov/sites/default/files/microsites/ostp/cairo-fact-sheet.pdf.

Organisation of Islamic Cooperation (OIC) (2011) *Current Stance of Science and Technology in OIC Countries*, October, [Online], Available: http://www.sesric.org/files/article/426.pdf.

Osama, A. (2011) *Islam Analysis, Science Vision 1441 Needs a Champion*, 11 January, [Online], Available: http://www.scidev.net/global/capacity-building/analysis-blog/islam-analysis-science-vision-1441-needs-a-champion-1.html.

Pew Research (2011) *Arab Spring Fails to Improve US Image*, 17 May, [Online], Available: http://www.pewglobal.org/2011/05/17/arab-spring-fails-to-improve-us-image/.

The Royal Society (2010) *A New Golden Age? The prospects for science and innovation in the Islamic world*, June, [Online], Available: https://royalsociety.org/~/media/Royal_Society_Content/policy/publications/2010/4294971224.pdf.

The Royal Society (2010) *New Frontiers in Science Diplomacy: Navigating the changing balance of power*, January, [Online], Available: https://royalsociety.org/~/media/Royal_Society_Content/policy/publications/2010/4294969468.pdf.

The Royal Society (2011) *Knowledge, Networks and Nations: Global scientific collaboration in the 21st century*, March, [Online], Available: https://royalsociety.org/~/media/Royal_Society_Content/policy/publications/2011/4294976134.pdf.

UNESCO (2010) *UNESCO Science Report 2010*, Paris: UNESCO Publishing.

US National Academy of Sciences (2014) *Arab-American Frontiers of Science, Engineering, and Medicine*, [Online], Available: http://sites.nationalacademies.org/PGA/dsc/AAFrontiers/PGA_081857.

US State Department (2012) *US Science Envoys Announced*, 8 November, [Online], Available: http://www.state.gov/r/pa/prs/ps/2012/11/200356.htm.

CHAPTER 3

Managing Plant Genetic Resources for Food and Agriculture: International Efforts and Lessons from the New Zealand Experience

Sarah Macindoe

With the world's population set to exceed nine billion by 2050, securing global food security into the future is fast becoming one of the most pressing challenges on the international agenda. Crop genetic resources are increasingly understood to be integral to combating this challenge, and consequently interest in the erosion of biodiversity and the related loss of plant genetic variation has grown significantly in recent years. Given the enormous interdependence of both countries and generations on this genetic diversity, its conservation and sustainable use are of global concern and have profound implications. This chapter seeks to assess the current state of international efforts to manage plant genetic resources for food and agriculture (PGRFA), and to explore some of the attendant challenges through a closer look at the New Zealand experience. The protection of PGRFA is reliant on technical and scientific innovation, collaboration and expertise, and thus presents a clear opportunity to operationalise the science diplomacy framework. As a small country heavily dependent on 'soft' power and deeply invested in the outcome of international negotiations and efforts in this area, New Zealand has a chance to harness science diplomacy in a way that is beneficial for all — both by contributing to global efforts to manage and conserve PGRFA and by using our related scientific skills and experience to advance our diplomatic interests overseas. However, Wellington's ability to do so is inherently shaped by a range of political, economic, legal and cultural considerations. Given the vital importance of PGRFA for ensuring future food production in the face of fundamental environmental and demographic changes, understanding the potential of science diplomacy in this area — and also its limitations — is an important step forward.

PGRFA: A Resource of Global Importance

Hunger and malnutrition affect more than 800 million globally, with over one in seven unable to gain sufficient protein and energy from their daily diet (Godfray *et al.*, 2010). Over 15 million die annually as a result (Alcázar, 2005), and it is estimated that malnourishment and hunger-related conditions collectively account for more than 53 percent of childhood deaths worldwide (Gibson, 2012: 4). At the same time, population growth continues at a remarkable rate. The world's population is expected to reach 8.3 billion by 2030 and to exceed 9 billion by 2050, and recent studies by the World Bank suggest that this will necessitate an increase in global food production of more than 70 percent (Godfray *et al.*, 2010: 812; Alcázar, 2005: 946; FAO, 2011b[1]: paragraph 4). This increased demand, however, must be met within the constraints imposed by the increasing scarcity of finite resources such as arable land and the increasing salience of environmental and demographic challenges such as climate change and urbanisation. Modelling indicates, for instance, that yields from rain-dependent agriculture in parts of Africa could drop by up to 50 percent by 2020, severely compromising food security in these regions. As Ramirez-Villegas *et al.* argue, the 'efficient use of agricultural diversity and genetic resources of both crops and forages will be needed to maintain … food production in the face of future challenges under future conditions' (Ramirez-Villegas *et al.*, 2013: 78–81).

2011–2020 has been christened the United Nations 'Decade on Biodiversity' in recognition of the critical role that agricultural biodiversity — of which PGRFA are a key component — plays as the biological basis for world food security, and to promote the need to conserve these resources (Morgera, 2012). The term PGRFA encompasses all 'genetic material of plant origin, including reproductive and vegetative propagating material of actual or potential value for food and agriculture' (Schaffrin *et al.*, 2006: 6). Genetic

[1] FAO (2011b) report hereafter referred to as CGRFA-13/11/Report.

variation enables crops and forages to adapt to biotic stresses, such as ever-evolving pests and diseases, and to overcome the constraints imposed by increasingly extreme climatic conditions, including drought, heat, flooding and salinity (FAO, n.d.b; Schaffrin *et al.*, 2006: 6; Ramirez-Villegas *et al.*, 2013: 81; Ramirez, 2013: 39). In other words, plant genetic resources (PGR) constitute the raw material that underpins attempts to improve existing crops and develop new varieties according to our future needs. By utilising genes from existing plant varieties, breeders and farmers are able, using either conventional selective breeding techniques or new genetic biotechnologies, to introduce novel and desirable traits — such as drought-tolerance or parasite-resistance — into crops important for food security at the local, regional and even global level (Schaffrin *et al.*, 2006: 6; Ramirez *et al.*, 2013: 39).

Clearly, PGRFA are critical for meeting human needs for food, health and economic security, and are thus immensely valuable. Despite growing international awareness of their importance, however, these resources continue to be depleted at an alarming rate. Approximately 90 percent of global crop varieties have been lost from farmers' fields since the early twentieth century, and losses continue to run at nearly two percent per year (FAO, n.d.b; Schaffrin *et al.*, 2006: 8). This dramatic reduction in plant genetic diversity can be attributed to a number of trends, including accelerating urbanisation, population pressures, changing dietary habits, loss of natural habitats, ecological degradation, climate change, overexploitation, and even legislative and policy changes (FAO, 2011b: paragraphs 9–11; Alcázar, 2005: 947; FAO, n.d.b). Most important, however, is the impact of modern agricultural systems and practices. Rising consumer demand for cheap food of uniform and predictable quality has sparked a move towards monocropping, or the introduction of standard, high-quality, homogenous plants at the expense of traditional, heterogeneous varieties, which has markedly shrunk the genetic pool (FAO, 2011b: paragraphs 9–11; Alcázar, 2005: 947). As Schaffrin, Görlach and Gerstetter argue, while modern agricultural methods have made an important contribution to raising crop yields, contemporary agriculture's tendency towards monoculture, global seed production and distribution, the move from traditional agricultural crops to planting cash crops for export, the abandonment of traditional farming practices and heavy reliance on a small group of core crops has had dire effects on agricultural biodiversity and the diversity of PGRFA' (Schaffrin *et al.*, 2006: 8).

The loss of plant genetic variation will have an impact on both developed and developing countries. Rapid globalisation and deepening economic integration are increasing the level of interdependence among nations, and currently all countries depend heavily on PGRFA originating in other regions of the world for their food production and sustainable agricultural development.[2] No country is self-sufficient for crop genetics resources: all are net recipients of foreign germplasm, and studies have shown that the average degree of genetic interdependence among countries for their most important crops is around 70 percent.[3] Given this interdependence, managing the conservation and sustainable use of plant genetic resources is a global concern, and efforts in this area must involve international cooperation. The concept of 'diplomacy for science' (or the enhancement of scientific understanding and promotion of scientific progress and innovation by way of mutually-beneficial international collaboration) is therefore of real significance. With the evolution of intellectual property laws and the establishment of state sovereignty over genetic resources — a development that will be discussed in more detail shortly — the need for countries to recognise the importance of cooperation and negotiation in adequately protecting PGRFA and driving the scientific advances these resources make possible is of fundamental importance. The advantages of cooperation for germplasm improvement and exchange, and the need for international sharing of genetic resources, have been demonstrated repeatedly, most recently in response to the outbreak and spread of wheat stem rust in Africa and West Asia.[4]

PGRFA and the Concept of the Global Commons

In many ways, the scope and nature of international cooperative efforts in the field of PGRFA are shaped by ideas regarding the ownership of genetic resources. Traditionally — or until the end of the twentieth century — plant genetic resources were conceived, at an international level, as being a

[2]On this point, see Alcázar (2005), Schaffrin *et al.* (2006: 7), Ramirez *et al.* (2013: 39–40).

[3]This figure is greater than 90 percent for Europe, and reaches almost 100 percent for North America and Australia. See Alcázar (2005: 949), and Schraffin *et al.* (2006: 7).

[4]For more detailed information, and more examples, see Ramirez *et al.* (2013: 40) and Alcázar (2005: 948).

'common heritage of mankind' (Roa-Rodriguez and van Dooren, 2008: 176). However, while such an appellative is still employed by advocacy groups such as Practical Action (who argue that 'the Earth's environmental resources are a common heritage of all mankind, which should be held in shared trust for a common future') (Practical Action, n.d.b), Roa-Rodriguez and van Dooren (2008: 178) explain that in its full international legal sense, the term 'common heritage' implies the presence of important additional elements, above and beyond those associated with the notion of common property, that have never played a part in the international regulation of PGR. As such, they argue that the term 'global commons' is more appropriate.

Halewood, López Noriega and Louafi describe global commons as 'a resource or resources shared by a group of people that is subject to social dilemmas' (Halewood *et al.*, 2013: 9). Commons are 'not exclusively subject to either state or private/market controls', but are rather cooperatively managed through the collective actions of all interested parties or stakeholders (Halewood *et al.*, 2013: 9). Unlike in other common property regimes (in which members regulate access to and use of a resource and have the right to exclude non-members), non-members of global commons are theoretically non-existent (Roa-Rodriguez and van Dooren, 2008: 178). While such 'all-inclusive' commons rarely apply to tangible resources — the subject of much traditional scholarship in this area — they are increasingly emerging in relation to intangible, or informational, resources. PGR provide an interesting lens through which to observe the evolution in commons thinking, as they are both tangible (i.e., sexual and vegetative seeds containing genetic material of actual of potential value) and intangible (i.e., germplasm contains the genetic code or 'blueprint', and is thus primarily an informational resource) in nature (Roa-Rodriguez and van Dooren, 2008: 179).

Until recently, scholarship has focussed on the 'traditional commons' — natural resources that are both rivalrous (in that one person's use detracts from others' ability to benefit from them) and non-excludable (in the sense that they are essentially open and available to all) (Halewood *et al.*, 2013: 9). However, the fact that human intervention (e.g. through seed-selection and cross-breeding) is required to ensure the conservation and development of genetic resources marks then them as 'new' commons, and distinguishes them from the natural resources of traditional commons theory. Furthermore, PGRFA are idiosyncratic in that *underuse*, rather than overuse, is the primary driver behind their erosion

(Halewood *et al.*, 2013: 10). In other words, as informational resources, PGR are fundamentally both non-excludable (although, of course, differing circumstances — such as access to wealth and knowledge — can render some better able than others to exploit them) and non-rivalrous, and thus do not suffer from 'tragic overuse' in the manner of many of their counterparts.

Despite this, however, over the past four decades national governments have increasingly demonstrated a willingness to expend considerable financial resources in order to exclude others from utilising various subsets of PGRFA. Such efforts employ both technological means — such as the development of infertile hybrids and the use of genetic use restriction technologies (GURTs) — and legal means — primarily the introduction of national and international intellectual property (IP) and access and benefit-sharing (ABS) laws (Halewood *et al.*, 2013: 10; Practical Action, n.d.b). As PGR continue to appreciate in value, both as objects of IP in their own right and as 'inputs into future innovation and product development' (Roa-Rodriguez and van Dooren, 2008: 179), pressure mounts for their privatisation. The result has been the transformation of the PGR commons from a resource that is open and available to one that is able to be commercially appropriated in a way that limits, or even prohibits, the use and access rights of others. Such evolution in the global community's conception of genetic resource 'ownership' is reflected in a number of international agreements related to the management and regulation of PGRFA. While the first comprehensive international agreement dealing with PGRFA, the 1983 International Undertaking on Plant Genetic Resources for Food and Agriculture (IU), refers to PGRFA as a common heritage, Resolution 3/91 (FAO, 2011b) qualifies this by affirming the sovereign rights of nations over their genetic resources. As such, the IU represents the beginning of 'a new era of PGR ownership and management' (Roa-Rodriguez and van Dooren, 2008: 185). Continuing this trend, the 1993 Convention on Biological Diversity (CBD) then established a new property domain in PGR at the international level. Genetic resources were repositioned as the property of sovereign states, with states afforded the authority to determine access to 'their' resources and to determine the conditions associated with their use by others (Roa-Rodriguez and van Dooren, 2008: 186).

As such, the context in which international efforts to manage PGRFA must operate within is characterised by state sovereignty over resources on which all nations depend. A well-functioning science diplomacy framework thus appears to be of critical importance. Improved awareness and understanding of the vital importance of PGR for future health and food security is needed in order to impress upon countries the need for cooperation in this area (science in diplomacy), and skilful diplomacy will be central to efforts to advance global collaboration and foster new partnerships — in both the public and private sectors — geared towards promoting their conservation and sustainable use (diplomacy for science). It is to the role of science and diplomacy in managing PGRFA at the international level that we now turn.

PGRFA, Science and Technology, and the International Policy Environment

The need to protect the integrity and diversity of PGRFA has been increasingly recognised both within the international policy arena and by the global science and technology community.

Protecting PGRFA: The role of science and technology

The Second Global Plan of Action for Plant Genetic Resources for Food and Agriculture, released in July 2011, emphasises the need to 'enhance linkages between scientific and technological innovations and their application to the conservation and use of PRGFA' (FAO, 2011b: paragraph 19), and points to a number of exciting new opportunities to employ emerging information and communication technologies and modern biotechnologies in the management of these resources.

Powerful new biotechnologies have increased the value of diverse PGRFA as donors of useful agricultural traits, and offer novel strategies for improving the efficacy and effectiveness of international efforts to protect these resources (Lidder and Sonnino, 2012: 6). The conservation of PGR is not enough: proper characterisation, evaluation, documentation and cataloguing of crop genetic resources are needed to allow their effective use. As Esquinas Alcázar explains, 'molecular genetics, genomics, proteomics, cryopreservation and ecogeographical remote-sensing technologies ... have greatly expanded the technological basis for the location, conservation and management of [PGRFA]' (Alcázar, 2005: 950), and the speed and cost at which genomes can now be sequenced means that such techniques can be more easily applied to the development of crop varieties that will respond well to environmental and biotic pressures (Godfray *et al.*, 2010: 815). Similarly, biotechnology applications, such as molecular markers (RFLPs; SNPs etc.) and genomic technologies, have made significant contributions to the characterisation of genetic resources (Lidder and Sonnino, 2012: 8), opening up 'completely new horizons for the exploitation of genetic diversity' (FAO, 2011b: paragraph 12). Advances in informatics and information and exchange technologies have also markedly improved our capacity to use, analyse and communicate related data and information (FAO, 2011b: paragraph 12).

Clearly, continuing to support scientific and technological progress for the protection and utilisation of PGRFA is an international priority. The generation and adaptation of biotechnologies requires 'a consistent level of financial and human resources and appropriate policies' (Lidder and Sonnino, 2012: 6). Finding ways to incentivise broad access and sustainability, whilst still encouraging 'a competitive and innovative private sector to make the best use of developing technology', is a major global governance challenge (Godfray *et al.*, 2010: 815). As Rao (2004: 143) argues,

> While there is a pressing need to ensure that available technologies are made accessible to a wider range of users through improved training and other capacity building initiatives, the existing technologies are also expensive, and given that most of the crop diversity is to be found in developing countries, the issue of resources assumes importance. Hence there is real need to maximise synergy through appropriate collaboration between various national, sub-regional and international levels, including sharing burdens and responsibilities, in order to use these technologies for effective conservation and use of plant genetic resources.

One major new initiative is the Generation Challenge Program (GCP), a consortium of national and international research institutes representing more than 200 partners in over 50 countries. Founded by the Consultative Group on International Agricultural Research (CGIAR) in 2003, the GCP aims to 'assist developing-world researchers to access technologies and to tap into a broader and richer pool of plant genetic diversity', and 'to use genetic diversity and advanced plant science to improve crops by adding value to breeding for drought-prone and harsh environments' (Generation Challenge Program, n.d.). The formation and success of such multi-national initiatives is a clear example of the science diplomacy framework in action. Similar efforts have been made to promote PGRFA within the international policy agenda.

Protecting PGRFA: International policy and negotiations

While the collective pooling of genetic resources is not new, the move towards their political and legal regulation is a more recent phenomenon. Efforts to formalise the conservation and pooling of genetic resources at the international level began with the 1967 technical conference on PGR, hosted by the Food and Agriculture Organization of the United Nations (FAO) and the International Biological Program, and the creation of CGIAR in 1971 (Halewood *et al.*, 2013: 4). Interest in PGRFA grew steadily throughout this time, and these resources became the subject of global regulation and concern. As interest in PGRFA deepened, however, so too did the desire by many countries to capitalise on their commercial value. The 1993 CBD constituted a landmark agreement in that it was the first international treaty to connect access to genetic resources to the fair and equitable sharing of the benefits arising from their utilisation, and yet it remains firmly grounded on the principle of national sovereignty over PGR (Schaffrin *et al.*, 2006: 13).

By the late 1990s, access to and use of PGRFA had been restricted by both sovereignty and IP claims. The introduction of intellectual property rights (IPRs) for new crop varieties and their associated genetic material in developed countries was quickly followed by the reassertion of national sovereignty and the application of clear restrictions on access to PGR in developing nations (Alcázar, 2005: 949; Roa-Rodriguez and van Dooren, 2008: 188). Rather than being available to all as a common resource of mankind, access to PGRFA increasingly became the subject of bilateral negotiations between states keen to exploit their natural resources. However, as Schaffrin *et al.* (2006: 10) argue, the 'international' nature of PGRFA has made such a bilateral approach seem inappropriate. PGRFA have been developed through the exchange of genetic material over both space and time, and as such it is often difficult to establish a sole nation of origin. It is also nearly impossible to identify the sovereign body authorised to provide access to them. The need to improve international cooperation and facilitate intergovernmental negotiations surrounding access to and use of PGRFA is therefore pressing.

Easily the most important recent development in internationally coordinated efforts to manage PGRFA has been the adoption of the International treaty on Plant Genetic Resources for Food and Agriculture (ITPGRFA) (FAO, 2011a). First adopted at the November 2001 FAO conference, the treaty entered into force on 29 June 2004 and constitutes the first legally binding international agreement focused on the conservation and sustainable use of PGRFA and the fair and equitable sharing of the benefits arising from their use (FAO, 2011a: 17); Halewood, 2013: 7). Of particular importance has been the creation of the multilateral system of access and benefit-sharing, through which the treaty facilitates no-cost access for farmers, breeders and scientists (from ratifying nations) to 64 of the most important agricultural crops for research, breeding and training purposes (FAO, 2011a: 105). The ITPGRFA also aims to ensure the sharing of benefits from the use of PGR through information-exchange initiatives, capacity-building and the transfer of technology (FAO, 2011b: paragraph 1; FAO, n.d.a). In more recent years, international attention has shifted towards negotiations surrounding the 2010 Nagoya Protocol on Access to Genetic Resources and the Fair and Equitable Sharing of Benefits Arising from their Utilization. Adopted in 2010, the Protocol aims to make operational the CBD's third objective (namely the fair and equitable sharing of the benefits arising out of the utilization of genetic resources') and to establish rules governing access to these resources and any associated traditional knowledge (NZMFAT, 2011).

Although a number of weaknesses persist, the ITPGRFA stands at the crossroads of trade, agriculture, and the environment,[5] and represents a dramatic step forward in the fight to protect biodiversity and promote global food security. It is the result of over 20 years of debate, including seven years of formal negotiations involving more than 160 members of the Commission on Genetic Resources for Food and Agriculture (CGRFA) as well as representatives from both the private sector and various non-governmental organisations (Alcázar, 2005: 950). However, while the treaty has enjoyed 'one of the fastest rates of national ratification of any international environmental

[5]This point is made by Alcázar (2005: 949) and Practical Action (n.d.a).

treaty', and currently boasts 127 members (Halewood *et al.*, 2013: 7), nine years have now passed since it entered into force and many actors are demonstrating a continued reluctance to fully engage. There are a number of reasons why this may be the case, principally among them the fact that the ITPGRFA and the Nagoya Protocol are as much economic as environmental agreements. Similarly, given the degree of international interdependence with regards to PGRFA, such negotiations have 'significant implications for countries involved in research and development using foreign genetic resources' (NZMFAT, 2011).

While PGRFA are resources of global importance, efforts to reach a global consensus regarding their protection and management are shaped by numerous political, legal, economic, scientific and cultural considerations, the salience of which vary across different nations and governments. Skilful diplomacy that acknowledges the inherent complexity of the issue and works to construct effective partnerships between actors with divergent needs and priorities for the protection and management of PGRFA will therefore take on an increasing importance. As a small country heavily dependent on soft power and deeply invested in the outcome of international efforts in this area, New Zealand has an opportunity to harness science diplomacy in a way that is beneficial for all — both by contributing to global efforts to manage and conserve PGRFA and by using our related scientific skills and experience to strengthen our international relationships and advance our diplomatic interests overseas. However, while science diplomacy is a framework of considerable potential value for New Zealand, it is not an approach completely devoid of challenges or constraints. Our ability to pursue both 'science for diplomacy' and 'diplomacy for science' in the context of PGRFA management is shaped by a range of political, economic, legal and cultural factors, all of which will play a central role in the formation of New Zealand policy and strategy in this field.

Managing PGRFA: The New Zealand Experience

New Zealand and science diplomacy

New Zealand is well-regarded internationally as an independent nation with a remarkably global and forward-leaning foreign policy (McLay, 2013a). As a small state with limited military and economic capabilities, it relies heavily on the use of 'soft' power and the smooth functioning of a rules-based international system to protect and advance its national interests. Consequently, 'New Zealand has long been committed to international cooperation — to strong bilateral, multilateral and regional relationships, to long-term sharing and to working with others on regional and global issues' (McLay, 2013a). Similarly, small advanced countries tend to be particularly flexible and thus able to adapt to and engage with international trends and opportunities. They are often characterised by well-developed science and innovation systems, and in many ways are pilots for the use of science and technology for economic and social advancement (Gluckman *et al.*, 2010).

In many ways, New Zealand is a world leader when it comes to scientific and technological innovation, in the agricultural and biotechnological sectors specifically. This presents a unique opportunity to foster beneficial new partnerships with regions of increasing strategic importance today, such as the Asia-Pacific, Africa and the Middle East. As developing countries grow and diversify, they are increasingly looking for new, long-term partnerships to deliver them the skills and technology they need, particularly in areas such as fisheries, agriculture and food security (McLay, 2013a). New Zealand's scientific expertise in these areas is internationally renowned, and sharing this expertise with others through channels such as the 'NZ Inc.' strategies thus presents an avenue through which scientific collaboration can be used to foster new connections and expand the country's diplomatic footprint (McLay, 2013a). Clearly, New Zealand as a nation is particularly well-suited to a 'science for diplomacy' approach to foreign relations. At the same time, much of its scientific success depends on access to resources sourced from overseas, and there is therefore 'a need to develop synergistic relationships with foreign partners' (Gluckman *et al.*, 2012). The advancement of New Zealand's scientific capabilities thus cannot be separated from the broader diplomatic agenda.

New Zealand and PGRFA: Challenges and opportunities

New Zealand's contribution to the conservation and sustainable use of PGRFA is an interesting lens through which to explore the opportunities and constraints associated with operationalising the science diplomacy framework.

Well-known as a hotspot of unique biodiversity, there is significant interest, both domestically and internationally, in accessing New Zealand's genetic resources — and often the associated traditional knowledge or mātauranga Māori (Wai262, 2011) — for research and commercial purposes (NZMFAT, 2011a; 2011b). These resources thus represent significant economic opportunities, as well as being a central component of the nation's brand. In other words, '[New Zealand's] indigenous biodiversity — [its] native species, their genetic diversity, and the habitats and ecosystems that support them — is of huge value to [the country] and its citizens; to [its] economy, [its] quality of life, and [its] sense of identity as a nation' (New Zealand Biotechnology, n.d.). It is therefore unsurprising that the decline on New Zealand's indigenous biodiversity has been described as the country's 'most pervasive environmental issue' (New Zealand Biodiversity, 2000).

New Zealand's interest in PGR goes beyond the need to conserve its indigenous genetic diversity, however. The country's primary production economy, particularly the agricultural, horticultural and forestry sectors, is crucially dependent on the genetic resources of introduced species (NZMFAT, 2009): land-based production is supported by fewer than 50 species of plants and animals — most of which are exotic — and current productivity in the primary industries (often the result of selective breeding) is enabled by access to a broad pool of introduced genetic diversity (New Zealand Biodiversity, 2000; FAO, 2007). Foreign genetic resources also constitute the backbone of New Zealand's burgeoning biotechnology industry, and some taonga species (e.g., kumara) are based on foreign PGR (New Zealand Biodiverisity, n.d.).

As John Allen, current CEO of New Zealand's Ministry of Foreign Affairs and Trade (MFAT) notes, the quality of New Zealand's agriculture (including its seeds and food and forage crops) drives Wellington's positioning on the world stage.[6] More importantly, genetic science in particular lies at the heart of both New Zealand's seed and pasture development, and therefore the genetic variation contained in plant genetic resources — both introduced and indigenous — is of vital importance to the country's agricultural success (New Zealand Biodiversity, 2000). Together, strong global

[6]John Allen, discussion with author, 11 February 2013, as part of the POLS231 class at the University of Otago, Dunedin, New Zealand.

interest in New Zealand's endemic genetic resources and the need to ensure continued access to PGR located overseas mean that the country's interests lie in an international regime that 'supports [its] ability to harness the benefits of being a provider of genetic resources' while also 'protecting [its] interests as a major user of foreign genetic resources' (NZMFAT, 2011a; 2011b). Similarly, maintaining the genetic diversity of important introduced species internationally is critical for their ongoing resilience to environmental change and their usefulness for our primary industries. As is explained in the New Zealand Biodiversity (2000) Strategy,

> Access to the gene pool overseas, or the maintenance of the diversity of genetic material of important production species within NZ, is crucial to manage risks to our economy and to maintain the potential for new economic activities in the future. New material needs to be able to be accessed to remain competitive in changing markets and in new biologically-based industries.

However, while the government historically invested heavily in the conservation of biodiversity, recent decades have instead seen an increasing reliance on market forces to conserve [the] genetic diversity of species important to production' (New Zealand Biodiversity, 2000) As such, those species for which the current market demand is less are seldom maintained. This loss of genetic diversity could have significant economic effects in the future, and finding ways to incentivise broader conservation efforts is therefore a key challenge for New Zealand with regards to PGR. While the task is substantial, New Zealand has a strong interest in biodiversity conservation and, as a major agricultural producer, 'is acutely aware of both the opportunities and challenges associated with the sustainable use of natural resources for agricultural development' (McLay, 2013a; 2013b). Additionally, New Zealand boasts world class science and technologies (notably genomic, reproductive and cloning), possesses extensive knowledge of the biology of important plants, grasses, trees and crops, and has access both to unique germplasm and to expanding genomic databases (New Zealand Biotechnology, n.d.).

Wellington therefore has both the interest and capability to play a driving role in the management of PGR, on both a domestic and international level. There are a number of benefits associated both with facilitating access to national genetic resources and ensuring continued access to those located internationally. However, using the genetic resources of indigenous species for commercial benefit 'raises ecological, commercial, cultural and ethical issues, as well as issues of access to genetic resources and how benefits of their use are shared' (New Zealand Biodiversity, 2000). The ability for New Zealand to utilise science diplomacy (particularly, 'diplomacy for science') with regards to genetic resource management is thus constrained, to some degree, by the need to find the 'sweet spot' between the national interest and the global interest.

New Zealand policy and PGRFA

Currently, national policy in the field of PGRFA is somewhat sparse. While New Zealand has been actively engaged in negotiations regarding both the ITPGRFA and the Nagoya Protocol, it has so far refrained from becoming a party to either, and continues to rely primarily on 'private agreements between plant breeders in dealing cooperatively with other countries' (FAO, 2007). As with all international negotiations, those regarding the treaty and the Nagoya Protocol were challenging, and encompassed a complex mix of scientific, political, legal, economic and indigenous rights issues. Observers noted an essential North-South divide between negotiating parties, with developed countries (wealthy but often lacking in diversity) seeking an international regime that 'facilitates access to genetic resources for their biotech industries and provides legal certainty regarding such access' and developing countries (poorer but often rich in diversity) demanding an 'all-encompassing and legally-binding regime that guarantees them a fair share of the benefits derived from the utilisation of their genetic resources'. New Zealand, which is both developed *and* diversity-rich, is thus unusually located in such discussions. A number of key considerations drive the country's negotiating position, and account for the failure to accede to the Protocol or the ITPGRFA:

1. *Commercial interests*: The global economic climate in recent years has shifted the government's attention toward maximising short- to medium-term trade benefits, and thus to prioritising applied science (science that can be commercialised and thus boost economic and private industry

growth) when it comes to resource investment decisions.[7] New Zealand's emerging biotechnology industry is of particular importance commercially. Revenue generated by the sector — primarily by way of intellectual property (IP) rights and patents for new technological and methodological inventions — reached NZ$351million in 2008–2009 (Wai262, 2011), and annual export earnings are expected to hit NZ$1billion in 2014 (New Zealand Biotechnology, n.d.). New Zealand thus has a keen interest in promoting the economic potential of this sector, and hence in protecting its sovereignty over its PGR by way of a robust IP regime. Related to this is the desire to maintain an internationally-competitive research environment, which necessitates both incentivising continued investment (by granting IPRs) and limiting the costs and restrictions imposed on research and innovation surrounding PGRFA (NZMFAT, 2009). The need to protect the integrity of its biotechnology industry and research infrastructure more generally is a key consideration for Wellington in determining its future role in the management of these resources.

2. *Obligations under the Treaty of Waitangi and the Wai262 Report:* In light of customary and Treaty rights (namely tino rangitiratanga and kaitiakitanga) granted to Māori in the Treaty of Waitangi and upheld in the Treaty of Waitangi Act 1975, New Zealand has an obligation to ensure that any domestic or international policy adopted regarding PGR contains 'appropriate recognition of the relationship between access and benefit-sharing activities and mātauranga Māori or traditional knowledge associated with genetic resources' (Wai262, 2011; NZMFAT, 2011). This is an issue of tremendous importance and consequence for Māori, given their historic status as kaitiaki (cultural guardians) of their taonga species. The kaitiakitanga and rangitiratanga rights of iwi regarding indigenous biological resources and associated mātauranga Māori was the subject of the Wai262 claim, lodged with the Waitangi Tribunal in 1991 (Wai262, 2011: xxii). Uncertainty regarding the findings of the Wai262 report has previously been a key obstacle to New Zealand's

[7]This point was made by both John Allen and Sir Peter Gluckman in private conversations with the author as part of the POLS231 class at the University of Otago, Dunedin, New Zealand, January–February 2013.

ratification of international agreements relating to PGRFA — particularly those with access and benefit-sharing components.

With the release of the final report in 2011, however, much of this uncertainty has been eliminated. The report determines that iwi and hapū do not have ownership over taonga species or related traditional knowledge, but affirms that 'their relationships with those species and associated knowledge are entitled to a reasonable degree of protection' (Wai262, 2011). Consequently, the Crown is entitled to implement laws and policies relating to both research into, and commercialisation of, New Zealand's plant genetic resources, but in doing so must, to the greatest extent possible, 'protect the authority of iwi and hapū in relation to their taonga species, so they can fulfil their obligations as kaitiaki'. As such, when it comes to negotiating access to New Zealand's indigenous PGR, New Zealand must consider the cultural interests of Māori and other local communities, as well as the fact that the approach it takes to approving the use of its own genetic resources will likely 'have important implications for reciprocal access to the indigenous genetic resources of other countries' (New Zealand Biodiversity, 2000).

3. *Legal/Policy Constraints*: A particular challenge for New Zealand during negotiations related to PGR has been the lack of an overarching domestic bioprospecting[8] policy to guide its negotiating position (NZMFAT, 2009). Any development in the country's approach to the management of genetic resources, however, must remain consistent with its international commitments, such as our membership of the 1994 WTO-TRIPS Agreement.[9]

In sum, New Zealand's objective regarding an international regime for the management of PGRFA is a practical and effective set of rules that will allow it to: (a) balance its interests as both a user and provider of genetic resources; (b) give appropriate recognition to the relationship between access and benefit-sharing activities and mātauranga Māori and maintain the

[8] Bioprospecting is 'the process of accessing and utilizing genetic resources to develop potentially useful products' (NZMED, 2007).

[9] The World Trade Organisation TRIPS Agreement (Agreement on Trade-Related Aspects of Intellectual Property Rights) is the most comprehensive global regulation of intellectual property, and covers both patents and plant variety rights. See Wai262 (2011).

Crown's ability to meet its obligations under the Treaty of Waitangi; and (c) enable the government to preserve an internationally-competitive research environment (NZMFAT, 2011). Wellington's failure to become a party to either of the two most significant international agreements on PGR — particularly now that much of the uncertainty regarding the rights of Māori has been removed — appears primarily to be the result of commercial considerations. Unfortunately, this is neither surprising nor particularly promising, and suggests that there may be certain constraints that science diplomacy, however actively pursued, may struggle to overcome.

The international context today continues to be characterised by state sovereignty, and despite the dramatic rise in the number of international institutions based on shared goals and norms, pursuit of the national interest (of which economic growth is a key component) remains a driving force in global politics. This raises the question of whether science diplomacy will have to encroach somewhat on state sovereignty in order to be truly effective.

Thinking ahead: New Zealand, PGRFA, and the potential of science diplomacy

In many ways, New Zealand has a chance to harness science diplomacy in a way that is beneficial for all — both by contributing to global efforts to manage and conserve PGRFA and by using its related scientific skills and experience to advance its diplomatic interests overseas.

Not only does New Zealand boast one of the world's fastest-growing biotechnology industries, it also possesses a number of key strengths in PGR-related research and innovation, including strong intellectual capital, endemic germplasm, and extensive biological databases (New Zeland Biotechnology, n.d.). Notably, New Zealand is a world leader in modern genomics techniques and genomics-assisted conventional breeding, with Crown Research Institutes such as Plant and Food Ltd. making major advances in the discovery of new markers (for selective breeding) and enjoying access to cutting-edge technologies such as high-throughput qPCR and microarray facilities. These genomic technologies are contributing to New Zealand's 'knowledge of the genes and molecular mechanisms that control key consumer and agronomic characteristics in plants'

(New Zealand Institute for Plant and Food Research, n.d.), and thus to the country's development of propagation systems and new crop varieties better adapted to climate change and more highly resistant to important pests and diseases.

There is thus great potential for New Zealand to employ a 'science for diplomacy' approach internationally in the field of PGR management and conservation. Continuing to invest strongly in its world-class breeding and genomics programmes is an incredibly promising way for New Zealand to reach the domestic 'sweet spot' between science, business, and diplomacy. Such initiatives promise both to generate significant economic rewards and to make a worthwhile contribution internationally to the management of PGRFA, allowing Wellington to leverage its scientific and technological capacity in way that advances the nation's interests on the world stage. Increasing the rate of international scientific collaboration, including working with developing countries to investigate and improve crop varieties of significant importance to them, is a way to maximise the global benefits which accrue from scientific and technological capacity and expertise. For a small nation such as New Zealand, science-based cooperation possesses potential to project relevance, build relationships and connections, gain international respect, and advance economic growth.

However, when considered in light of New Zealand's role in the international management of PGR, the potential of 'diplomacy for science' appears less certain than that of 'science for diplomacy'. Many scientific challenges — including the global erosion of biodiversity and the related loss of genetic resources — are transnational in nature, and as such require a transnational or multilateral response. Given that countries almost invariably have different priorities and operate within different constraints, and that in many cases vary in their degree of international influence, incredibly skilful diplomacy is needed to foster agreement and compromise in the name of the common good. While this is possible, and while talented and dedicated diplomats make substantial gains more often than is commonly recognised, the continued centrality of sovereignty and national interest to the international system means that adequate compromises are rarely achieved. The repeated inability of states to make real progress on tackling climate change is a case in point. Similarly, New Zealand's reluctance to prioritise the common pooling of PGR over the current and future trade and economic benefits these resources represent — and its ability to determine who may and may not access 'its' PGR — indicates that the concept of state sovereignty may pose a significant challenge to some elements of science diplomacy.

As such, science diplomacy may have to encroach somewhat on sovereignty to be effective. This is not a totally novel idea. The emerging concept of the 'Responsibility to Protect' (R2P) is premised on a notion of conditional sovereignty, or the idea that sovereignty is dependent on the state being both willing and able to adequately protect its population from harm. A fully functioning science diplomacy framework may implicitly require a similar definitional and/or conceptual shift. In other words, for 'diplomacy for science' to achieve the desired ends regarding the conservation and sustainable use of PGRFA at the global level, a return to the classification of such resources as a global commons (or even common heritage) may be necessary. The success of diplomatic efforts aimed at advancing cooperation and progress on scientific issues of global importance may depend, to some extent, on states relinquishing their sovereign claim to resources on which they are interdependent. Such a reconceptualization of sovereignty, however, would have immense political and legal implications, and would likely face concerted resistance.

While it is common to talk of the potential promise of 'science diplomacy' more generally, it must therefore be acknowledged that the different components of the science diplomacy framework (science in diplomacy, science for diplomacy and diplomacy for science) rest on different premises and are subject to different constraints. Consequently, they are likely to have noticeably divergent chances of success in different contexts. A point so elemental it is often forgotten is that whereas the end goal of diplomacy for science is to advance scientific knowledge and collaboration, the end goal of science for diplomacy is to improve a nation's foreign relations. The former is (typically) driven by the global interest and may require some subversion of the national interest; the latter advances the national interest, often at the expense of an often vaguely-defined and conceptually-remote international interest. Diplomacy for science may therefore be an inherently more challenging pursuit.

Finding the 'sweet spot' between the national and global interest is a critically important task, but also an incredibly difficult one. This is not to diminish the potential of science diplomacy. The extent to which the science diplomacy framework is utilised in coming years will have a remarkable influence on the global trajectory through the twenty-first century. It will not, however, be able to reconfigure the international landscape singlehandedly.

References

Alcázar, J.E. (2005) 'Protecting crop genetic diversity for food security: political, ethical and technical challenges', *Nature Reviews Genetics,* 6: 946–953. DOI: 10.1038/nrg1729.

Food and Agriculture Organization of the United Nations (FAO) (n.d.a) 'Use them or lose them', *Commission on Genetic Resources for Food and Agriculture,* [Online], Available: http://www.fao.org/nr/cgrfa/cthemes/plants/en/ [1 February 2013].

——— (2007) 'New Zealand: Country Report on the State of Plant Genetic Resources for Food and Agriculture', *2nd Country Report by Commission on Genetic Resources for Food and Agriculture,* August 2007, [Online], Available: http://www.fao.org/docrep/013/i1500e/New%20Zealand.pdf [1 February 2013].

——— (2011a) 'Introduction to the International Treaty on Plant Genetic Resources for Food and Agriculture', *Food and Agriculture Organization of the United Nations,* 1st edition, Rome, Italy, February 2011, [Online], Available: http://www.planttreaty.org/sites/default/files/edm1_full_en.pdf [18 October 2013].

——— (2011b) 'Report of the Commission on Genetic Resources for Food and Agriculture', *Thirteenth Regular Session,* Rome, Italy, 18–22 July 2011, [Online], Available: http://www.fao.org/docrep/meeting/024/mc192e.pdf [1February 2013].

——— (2013) *Commission on Genetic Resources for Food and Agriculture,* [Online], Available: http://www.fao.org/nr/cgrfa/en/ [1 February 2013].

Generation Challenge Program (n.d.) 'Who We Are and History', *Generation Challenge Program,* [Online], Available: http://www.generationcp.org/about-us/who-we-are [12 November 2013].

Gibson, M. (2012) *The Feeding of Nations: Redefining food security for the 21st century.* Boca Raton, FL: CRC Press.

Gluckman, P.D., Goldson S. L. and Beedle A.S. (2012) 'How a small country can use science diplomacy: A view from New Zealand', *Science and Diplomacy,* 24 May, [Online], Available: http://www.sciencediplomacy.org/perspective/2012/how-small-country-can-use-science-diplomacy.

Godfray, H.C.J., Beddington J.R., Crute I.R., Haddard L., Lawrence D., Muir J.F., Pretty J., Robinson S., Thomas S.M. and Toulmin C. (2010) 'Food Security: The Challenge of Feeding 9 Billion People', *Science,* 327(5967): 812–818. DOI: 10.1126/science.1185383.

Halewood, M., López Noriega I. and Louafi S. (2013) 'The global crop commons and access and benefit-sharing laws', in Halewood M., López Noriega I. and Louafi S. (eds.) *Crop Genetic Resources as a Global Commons: Challenges in International Law and Governance*, New York: Routledge, pp. 1–36.

Wai262 (2011) 'Ko Aotearoa Tēnei: A Report into Claims Concerning New Zealand Law and Policy Affecting Māori Culture and Identity. Te Taumata tuatahi', Volume 2, *Waitangi Tribunal Report 2011,* Wellington, New Zealand: Legislation Direct, [Online], Available: https://forms.justice.govt.nz/search/WT/reports.html [1 February 2013].

Lidder, P. and Sonnino A. (2012) 'Biotechnologies for the management of genetic resources for food and agriculture', *Advances in Genetics,* 78: 1–167. DOI: 10.1016/B978-0-12-394394-1.00001-8.

McLay, J. (2013a) 'New Zealand and the United Nations: Small State, Big Challenge', *Lecture delivered to the University of Auckland Political Studies Department,* 27 August 2013, [Online], Available: http://www.nzunsc.govt.nz/docs/Jim-McLay-speech-Small-State-Big%20Challenge-Aug-13.pdf [18 October 2013].

——— (2013b) 'Second Committee of the 68th Session of the UN General Assembly Agenda item 25: Agricultural Development, Food Security and Nutrition', *Statement delivered 28 October 2013,* [Online], Available: http://www.mfat.govt.nz/Media-and-publications/Media?MFAT-speeches/2013/0-28-Oct-2013.php [3 November 2013].

Ministry of Economic Development (New Zealand) (NZMED) (2007) *Bioprospecting: Harnessing Benefits for New Zealand — A policy framework discussion,* [Online], Available: http://www.med.govt.nz/sectors-industries/natural-resources/pdf-docs-library/biodiscovery/bioprospecting-harnessing-benefits-for-new-zealand.pdf [23 January 2013].

Ministry of Foreign Affairs and Trade (New Zealand) (NZMFAT) (2009) 'Genetic Resources: Negotiations on an International Regime on "Access and Benefit-Sharing"', *Briefing Note Provided to the Chair, Cabinet External Relations and Defence Committee,* [Online], Available: http://www.mfat.govt.nz/downloads/global-issues/regulations-access-benefit-sharing.pdf [23 January 2013].

Ministry of Foreign Affairs and Trade (New Zealand) (NZMFAT) (2011) *Genetic Resources (Access and Benefit-Sharing),* 1 February, [Online], Available: http://www.mfat.govt.nz/Foreign-Relations/1-Global-Issues/Environment/7-Species-Conservation/geneticres.php [23 January 2013].

Morgera, E. (2012) 'Briefing Note of the Second High-Level Round Table on the International Treaty on Plant Genetic Resources for Food and Agriculture (ITPGR)', *International Institute for Sustainable Development,* 22 June, [Online], Available: http://www.iisd.ca/uncsd/rio20/itpgrfa [17 January 2013].

New Zealand Biodiversity (2000) *The New Zealand Biodiversity Strategy — Executive Summary,* [Online], Available: http://www.biodiversity.govt.nz/picture/doing/nzbs/summary.html [24 January 2013].

New Zealand Biotechnology (n.d.) *New Zealand Biotechnology Overview*, [Online], Available: www.nzbio.org.nz [24 January 2013].

New Zealand Institute for Plant and Food Research (n.d.) *Breeding and Genomics,* Available: http://www.plantandfood.co.nz [24 January 2013].

Practical Action (n.d.a) *International Seed Treaty on Major Food Crops*, [Online], Available: http://practicalaction.org/seed_treaty [24 January 2013].

——— (n.d.b) *Agricultural Biodiversity*, [Online], Available: http://practicalaction.org/global_commons [24 January 2013].

Ramirez, M., Ortiz R., Taba S., Sebastián L., Peralta E., Williams D.E., Ebert A. and Vézina A. (2013) 'Demonstrating Interdependence on Plant Genetic Resources for Food and Agriculture', in Halewood M., López Noreiga I. and Louafi S. (eds.) *Crop Genetic Resources as a Global Commons: Challenges in International Law and Governance,* New York: Routledge, pp. 39–61.

Ramirez-Villegas, J., Jarvis A., Fujisaka S., Hanson J. and Leibing C. (2013) 'Crop and Forage Genetic Resources: International Interdependence in the face of Climate Change', in Halewood M., López Noreiga I. and Louafi S. (eds.) *Crop Genetic Resources as a Global Commons: Challenges in International Law and Governance,* New York: Routledge, pp. 78–98.

Rao, N.K. (2004) 'Plant genetic resources: Advancing conservation and use through biotechnology', *African Journal of Biotechnology,* 3(2): 136–145, Available: http://hdl.handle.net/1807/3508.

Roa-Rodriguez, C. and van Dooren T. (2008) 'Shifting Common Spaces of Plant Genetic Resources in the International Regulation of Property', *Journal of World Intellectual Property*, 11(3): 176–202. DOI: 10.1111/j.1747-1796.2008.00342.x.

Schaffrin, D., Görlach B. and Gerstetter C. (2006) 'The International Treaty on Plant Genetic Resources for Food and Agriculture — Implications for Developing Countries and Interdependence with International Biodiversity and Intellectual Property Law', *Final Report IPDEV Work Package 5*, [Online], Available: http://www.ecologic.eu/download/projekte/1800-1849/1802/wp5_final_report.pdf [6 June 2014].

CHAPTER 4

Antarctic Science: A Case for Extending Diplomacy for Science

Gary Wilson

Discovery

At the turn of the 19th Century it was still not certain that the Antarctic continent existed. While James Cook is generally credited with the discovery of Antarctica in his 1774–1775 voyage, he did not actually sight the continent. In fact, the first sighting of the southern continent was by Thaddeus von Bellingshausen of the Russian Imperial Navy on 27 January 1820 (Bonner, 1987). However, it would take another 10–20 years before the true scale of the great southern continent started to emerge, motivated in parallel by commerce and science. The hunt for whales drew whaling fleets into the Southern Ocean and in the early 1830's John Biscoe — a whaling captain — also circumnavigated Antarctica (Jones, 1981).

Not long after these early circumnavigations, French and British naval expeditions sailed south to search for the south magnetic pole: these were the expeditions of Durmont D'Urville and James Clark Ross, respectively (Walton and Doake, 1987). Not that they found it — that was left to the Australian Edgeworth David on Shackleton's Nimrod expedition in 1909 (Tingey, 1983). Neither did Dumont D'Urville or James Clark Ross work together, but their expeditions probably marked the start of scientific undertakings in Antarctica.

By the early 20th Century the race was on to the South Pole. Robert Falcon Scott's first expedition (the Discovery Expedition) reached 82°12' South in 1902. Ernest Shackleton returned six years later and reached 88°23' South in January 1909, only 180 km from the pole before having to turn back. In January 1912, on the Terra Nova Expedition, Scott finally reached the Pole only to discover that Roald Amundsen had beaten him and reached the pole a month earlier in December 1911. Scott and his South party did not return to the hut at Cape Evans and perished on the Ross Ice Shelf only 11 miles (approx. 18 kilometres) short of 1 ton depot (Preston, 1999).

While earlier territorial claims had been laid for subantarctic islands, the claims for sectors of Antarctica began in parallel with the early expeditions to the pole (Berkman, 2001). Britain claimed the Ross Sea sector, which they later passed to New Zealand by order-in council in 1923. Following Shackleton's eventful Imperial Trans-Antarctic Expedition (1914–17), Britain also claimed the Antarctic Peninsula and Weddell Sea sector. The French claimed Terre Adélie (1924), named by Durmont D'Urville in his earlier visit, followed another decade later by another British claim to much of East Antarctica (later to pass to Australia) based on the British New Zealand Antarctic Research Expedition of 1929–1931 to Wilkes Land. Norway claimed the remaining sector of East Antarctica in 1939 followed by Chile in 1940 and Argentina in 1943 who made overlapping claims to the Antarctic Peninsula (Berkman, 2001). By 1943, seven nations had claimed segments of the Antarctic continent, three of them overlapping and so the interest at government level had begun. The United States reserved its right to make a claim (Belanger, 2006), leaving Marie Byrd Land as the only unclaimed sector of the continent (Figure 1).

Science

Through the race for the pole and territorial claims, science kept a foothold — Scott's Terra Nova expedition (1910–1913) had a significant science programme, particularly in Geology and Zoology. In fact, it was Scott's intention at the time that the expedition be remembered for its scientific effort as Raymond Priestly, a geologist and meteorologist on the Nimrod and Terra Nova expeditions later recalled in his famous quote (National Maritime Museum, 2000):

> Scott for Scientific method, Amundsen for speed and efficiency, but when disaster strikes and all hope is gone, get down on your knees and pray for Shackleton.

The Terra Nova Expedition included a large cohort of scientists led by Zoologist Edward Wilson. Among them were geologists Frank Debenham,

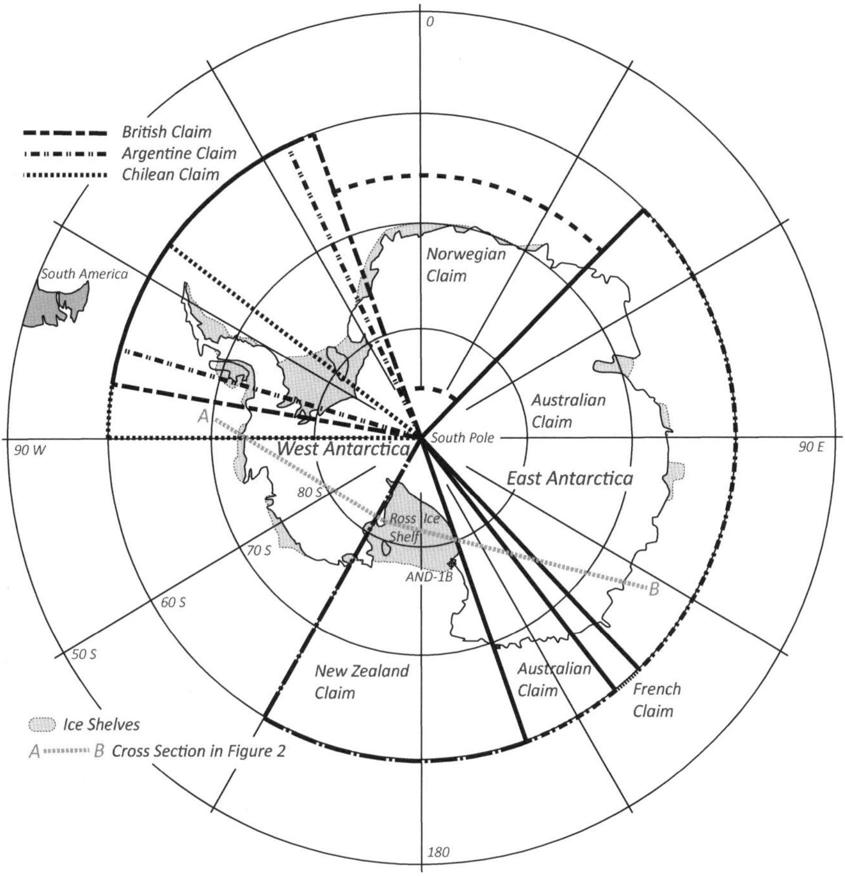

Figure 1. Antarctic territorial claims.

Raymond Priestley and Griffith Taylor. Griffith Taylor led the geological expeditions to southern Victoria Land (the Western Journeys) and Raymond Priestley was a member of the Northern Party, who wintered at Cape Adare (Huxley, 1913). Accompanying Wilson was Apsley Cherry-Gerrard, famous for his account of the Winter Journey to Cape Crozier ('The Worst Journey in the World'), to collect Emperor Penguin eggs in the hope of proving the link between reptiles and birds (Cherry-Gerrard, 1922).

Scientific exploration of the continent continued after the First World War with American explorers putting new technology to use. Particularly important were the aerial forays of Admiral Richard Byrd that opened up the continent beyond its coastlines and the early pole-bound explorers tracks.

Byrd undertook five expeditions to the Antarctic between 1929 and 1956. His third expedition (1939–1941) was the first to have official US government backing and include a programme of scientific exploration. His fourth expedition, the US Navy Antarctic Developments Program 1946–47, more commonly known as 'Operation Highjump', included a formal plan for scientific exploration. More importantly though, was the recognition that the scientific discoveries would be best obtained by a collaborative international scientific effort (Belanger, 2006).

It was the aspiration of international collaboration that led to the International Geophysical Year 1957–58 (IGY), a global programme of research, which also included an unprecedented programme of scientific exploration of Antarctica including scientists from 12 nations (Argentina, Australia, Belgium, Chile, France, Great Britain, Japan, New Zealand, Norway, South Africa, USSR, and the United States). IGY also saw the establishment of permanent scientific bases (Summerhayes, 2008) and the first transantarctic crossing by Vivian Fuchs (supported by Ed Hillary): a journey not attempted since Shackleton's Endurance Expedition in 1914. It had been Lloyd Berkner of Carnegie Institution of Washington and veteran of Admiral Byrd's expeditions that mooted the idea of a third polar year, which led to the IGY. Thus through the International Council of Scientific Unions (ICSU), Antarctic Science led the way in a global cooperation involving more than 60 countries who participated in the global efforts of IGY. Coordinated planning began in 1955 for the 12 nations that would participate in Antarctic Research. For the most part, ICSU managed to keep the focus on scientific planning and away from political aspirations. For this reason, geological research was omitted from the Antarctic IGY research programme, which was limited to the observation of geophysical, glaciological, atmospheric and ionospheric phenomena (Summerhayes, 2008).

The IGY expeditions mapped out the basic shape and nature of the Antarctic continent, its ice sheets, and its atmospheric and oceanographic properties. This gave the first indication that Antarctica was covered in 30 million-cubic kilometers of ice, which, if melted, would raise global sea levels by more than 60 metres (Bentley *et al.*, 1964). It also became clear that the Antarctic Ice Sheet is split in two by the Transantarctic Mountains — to the East it sits on the continent that would rest above sea level if the weight of the ice were removed, but to the west it sits below sea level pinned on a series of islands (Figure 2).

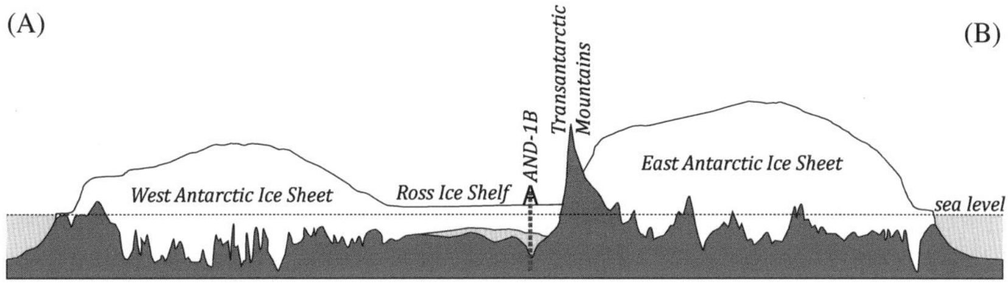

Figure 2. Cross section of Antarctica. See Figure 1 for line of cross section. Vertical exaggeration = 250 times.

Diplomacy

Diplomacy did have a small hand in helping the IGY off the ground. The USSR declared its intent to build several bases in the Antarctic, which helped the United States decide to expand its efforts and commit a greater level of funding (Belanger, 2006). After the success of the IGY, an extended research effort was proposed by the scientific participants, and again it was the Russians that spurred on other participants by declaring their intent to continue (Belanger, 2006). Thus in February, 1958, the first meeting of the Scientific Committee on Antarctic Research (SCAR) was held with the purpose of coordinating ongoing international scientific efforts in Antarctica (Summerhayes, 2008).

It was another year later, in 1959, that the Antarctic Treaty was drafted by the twelve nations that participated in the IGY, coming into effect on 23 June 1961 (Berkman, 2001). A further 35 nations have joined the treaty since its original signing. The treaty is very simple[1] and contains only 14 articles (Table 1), but it is the preamble[2] that recognises the significant role of science in its construct:

> Recognizing that it is in the interest of all mankind that Antarctica shall continue for ever to be used exclusively for peaceful purposes and shall not become the scene or object of international discord;
>
> Acknowledging the substantial contributions to scientific knowledge resulting from international cooperation in scientific investigation in Antarctica;

[1]The official website for the treaty can be found in the following link: http://ats.aq/.
[2]Specifically, found here: http://www.ats.aq/e/ats.htm.

Table 1. Summary of the Antarctic Treaty (1959).

Article 1	Antarctica shall be used for peaceful purposes only. Military presence is restricted to the support of scientific research
Article 2	Scientific Research and cooperation established in the IGY should continue
Article 3	There should be international cooperation in Scientific Research through exchange of personnel and knowledge
Article 4	The treaty does not recognise, dispute or endorse territorial claims and there shall be no new claims
Article 5	Antarctica will be Nuclear Free
Article 6	The treaty applies to the area south of 60° latitude
Article 7	Observers from each members shall be given free access to inspect the activities of other members
Article 8	Allows for each member state to have due restriction over their own nationals working in Antarctica
Article 9	States that there shall be frequent treaty meetings
Article 10	All members shall discourage any activities by any other nation that are contrary to this treaty
Article 11	Any dispute shall be settled peacefully and if that is not possible by the International Court
Articles 12, 13 and 14	Deal with how the treaty should be upheld, interpreted or modified

> Convinced that the establishment of a firm foundation for the continuation and development of such cooperation on the basis of freedom of scientific investigation in Antarctica as applied during the International Geophysical Year accords with the interests of science and the progress of all mankind;
>
> Convinced also that a treaty ensuring the use of Antarctica for peaceful purposes only and the continuance of international harmony in Antarctica will further the purposes and principles embodied in the Charter of the United Nations.

The first three articles deal with the focus of the treaty — science and cooperation. The next three articles deal with the scope of the treaty — south of 60°South, prohibition of nuclear explosions and disposal of radioactive waste, and that it would not supplant previous territorial claims. The remaining articles deal with enforcement and operational procedures.

The advisory role of SCAR to the treaty was adopted at the first consultative treaty meeting in 1961. SCAR was tasked with initiating and coordinating high quality international scientific research in the Antarctic region and on the role of the Antarctic region in the global earth system. Thus, science continued to play a key role in the international cooperation in Antarctica. Antarctic Treaty Meetings have been held every year and SCAR meetings every second year since the signing of the treaty in 1960.

SCAR is organized around Scientific Standing Groups in Geosciences, Life Sciences and Physical Sciences. In its first 30–40 years, SCAR was consumed with the rate of scientific discovery and focused on the exchange of information between treaty members and cooperation in these activities as intended by the treaty. Essentially, diplomacy drove the scientific discovery and cooperation (diplomacy for science). But in 2000, SCAR changed its focus to having science drive the collaboration through internationally coordinated Scientific Research Programmes (Summerhayes, 2008).

Funded by the national academies in member countries, SCAR is able to maintain independence of national aspirations in the Antarctic and — with its new structure — focus on fundamental scientific research requiring internationally collaborative effort. For example, Turner *et al.* (2009) produced a comprehensive synthesis report on Antarctic Climate Change and the Environment to mark the International Polar Year 2007–2008, which has led to new SCAR Scientific Research programmes concerned with impacts of changing climate on ice sheets, southern ocean and biota.[3]

A Global Environment

The report by Turner *et al.* (2009) and many other recent studies reinforce the fact that Antarctica plays a critical role in the global ocean and atmospheric system. With respect to atmosphere, Antarctica sets the hemispheric temperature gradient that, combined with the Coriolis effect from a rotating planet, defines atmospheric circulation and its subdivision into the Hadley, Ferrel and Polar cells from the equator to pole, respectively (Figure 3). Each circulates in an opposing direction around the Earth driving ocean surface currents and creating unstable fronts between cells.

---

[3] For the more information, see the official website: http://www.scar.org/researchgroups/.

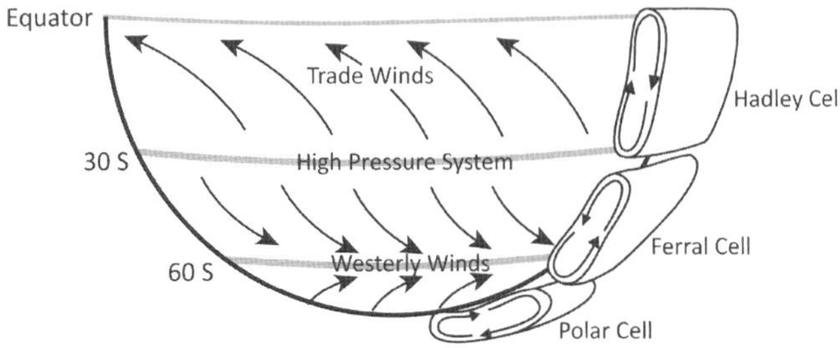

Figure 3. Southern Hemisphere atmospheric circulation.

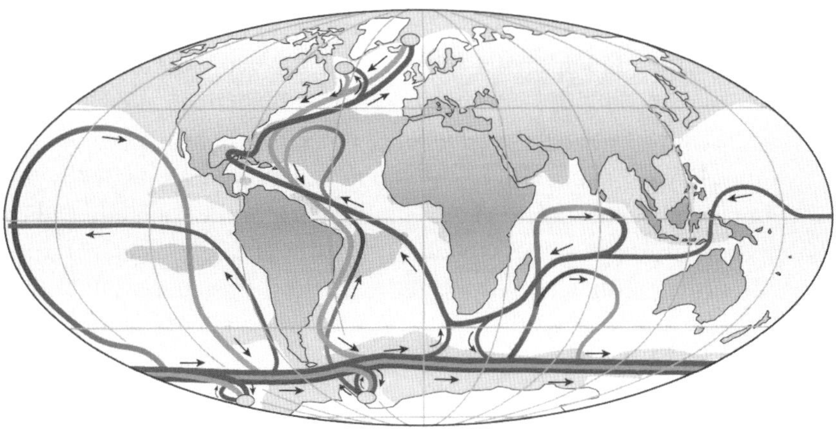

Figure 4. Schematic representation of global thermohaline circulation.

Notes: Dark lines = surface circulation, light lines = deep water circulation, circles = nodes of bottom water production.

In the Ocean, a similar subdivision occurs with divergent and convergent fronts between different water masses that progressively warm as one moves north from the Antarctic. Freezing ice on the surface of the Antarctic Ocean also produces Antarctic Bottom Water — supercooled, supersaline water that sinks and flows north along the ocean floor, driving global ocean circulation and the distribution of heat around the planet (Figure 4, taken from Rahmstorf, 2002). Deep water is produced predominantly in the Weddell and Ross seas and flows north along the western margins of each ocean to rise in the Northern Hemisphere and return south at the surface or intermediate depths (Rahmstorf, 2002).

A Warming Globe

The most recent report of the Intergovernmental Panel on Climate Change (IPCC, 2013) notes that the planet will continue to warm with increasing levels of atmospheric CO_2, which are expected to reach between 500 and 800 parts per million (henceforth ppm) by 2100 with a projected global average temperature increase of between 2 and 4°C. Most worryingly, the warming is likely to be 2–4 times greater at the poles than the equator. Warming of 8–10°C or more in the Antarctic is likely to have far reaching implications for the rest of the globe with respect to ocean and atmospheric circulation as well as rising sea levels from melting ice.

Understanding the degree to which future warming may affect Antarctica and its ice sheets will take more than sharing of the findings and data from individual Antarctic programmes: a multinational approach will be required to mount expeditions of the scale required.

The multinational ANDRILL (Antarctic Geological Drilling) project is an example of such. New Zealand had already been developing the technology and the international partnerships needed to undertake geological drilling in the Antarctic, but drilling beneath Antarctica's massive floating ice shelves — so as to see how often they had collapsed and why — required a new level of international collaboration. ANDRILL brought together scientists from New Zealand, the United States, Italy and Germany and the collective efforts of their national Antarctic programmes to drill through 86 metres of floating ice moving at approx. 1 metre per day, 860 metres of water and recover a 1285 metres core (the AND-1B core) from the sediments beneath the sea floor (Figure 5).

Like the IGY, the ANDRILL collaborative effort developed in several stages. Also like IGY, it took 10-years from conception to completion. Initially, the scientific rationale and partnership was brought together by a community based group, the ANDRILL Science Committee (ASC), funded by subscription from nations interested in developing the scientific programme. The ASC undertook feasibility studies of the technical requirements, coordinated site survey efforts and the development of an international proposal for drilling (Wilson *et al.*, 2012a). At this point the national Antarctic programmes of the member nations, the ANDRILL Operations Management Group (AOMG), negotiated a multinational agreement to provide the logistic support required by the scientific programme and then appointed a Science Implementation Committee, the McMurdo ANDRILL Science Implementation Committee

Figure 5. The ANDRILL (Antarctic Geological Drilling) rig on the McMurdo Ross Ice Shelf south of Mount Erebus. Inset shows schematic view of the AND-1B core recovery. *Note*: Artwork originally by Angie Fox. AND-1B location shown on Figures 1 and 2.

(MASIC), to ensure science plans took effect and that national interests were represented in the scientific staffing, that samples and data were correctly reposited and that the results were published in a timely manner.

As well as developing the combined riser and hot water drilling system to enable the recovery of sediment cores from beneath the McMurdo Ice Shelf, ANDRILL also developed a series of sub-ice imaging methodologies, a programme of modelling enabling a wider perspective to be developed in parallel with the results from the drilling, and a programme of outreach that brought the results into many fora.[4]

Results from Scientific Drilling in Antarctica: A New Challenge for Diplomacy

In May 2013, National Oceanic and Atmospheric Administration's (NOAA) Mauna Loa Observatory measured, for the first time, 400 ppm carbon dioxide

[4] For more information, see the official website: http://www.andrill.org.

in the atmosphere. Yet, mankind barely blinked. Why? What does 400 ppm really mean? The ANDRILL core from beneath the McMurdo Ice Shelf provides some answers; The AND-1B record samples the last time the Earth was warmer than today, by about 3°C on average, with average atmospheric carbon dioxide (CO_2) levels between 330 and 400 ppm (Seki *et al.*, 2010).

Interpreting the past environments from the AND-1B record is surprisingly straightforward at the broadest level. A pattern of recurring rock-types record ice grounded on the sea floor at the drill-site, a floating ice shelf, and open water conditions (Figure 6) (McKay *et al.*, 2010). The chronology (Wilson *et al.*, 2012b) shows that the sediment accumulated relatively rapidly punctuated by breaks in sediment accumulation or periods of erosion. This pattern of erosion from an advancing glacier grounded

(A) (B) (C) (D) (E)

Figure 6. Examples of sediment recovered from beneath the McMurdo Ross Ice Shelf in the AND-1B drill core.

Note: (A) grounded ice and erosion, (B) ice sheet grounding line, (C) ice shelf, (D) open Ross Sea with some ice bergs, and (E) warm open Ross Sea with no ice bergs.

on the sea floor followed by deposition beneath the glacier, rapid accumulation as the glacier melted and retreated and then biogenic deposition in open water (no ice-shelf), occurred more than 40 times over the past 5 million years, but was most pronounced between 3 and 5 million years ago (Naish *et al.*, 2009).

Computerised models which link ice sheets, ocean and atmosphere, predict that global conditions in the Pliocene Era (3–5 million years ago) as recorded in the AND-1B drillcore resulted in retreat and at times complete loss of the West Antarctic Ice Sheet on a regular basis (Figure 7, taken from Pollard and DeConto, 2009). In an earlier modelling study, DeConto and Pollard (2003) showed that the Antarctic ice sheets responded dramatically to elevated levels of atmospheric CO_2. Three-times pre-industrial levels of CO_2 (840 ppm) saw complete loss of Antarctic ice sheets but at two-times

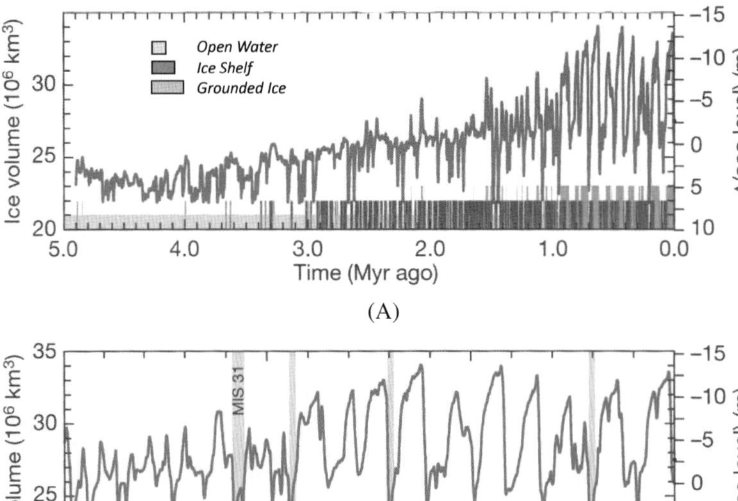

(A)

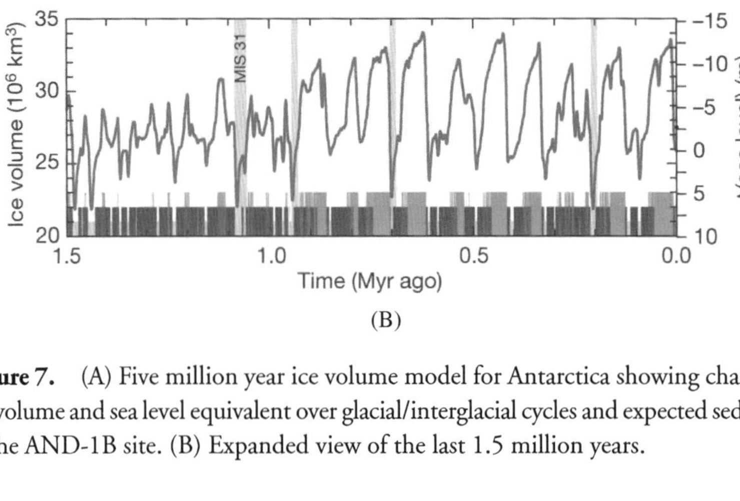

(B)

Figure 7. (A) Five million year ice volume model for Antarctica showing change in ice volume and sea level equivalent over glacial/interglacial cycles and expected sediment at the AND-1B site. (B) Expanded view of the last 1.5 million years.

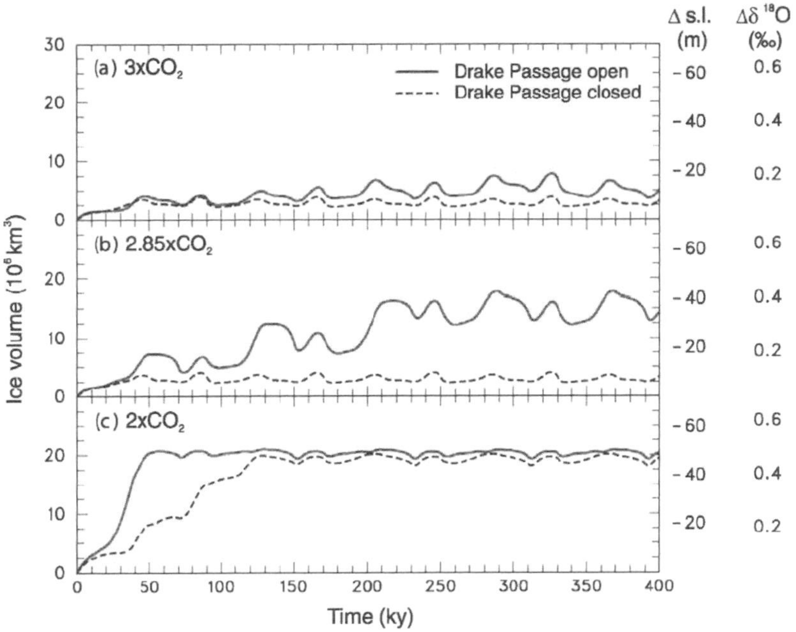

Figure 8. Modelled Antarctic ice sheet growth at difference levels of atmospheric CO_2.

preindustrial levels of CO_2 (560 ppm) saw modest change in ice sheets, but at levels between two and three times pre-industrial CO_2, ice sheets became highly responsive to other forcing agents such as changing levels of solar energy (Figure 8, taken from DeCOnto and Pollard, 2003). What the more recent ANDRILL study shows is that the marine based West Antarctic Ice Sheet responded to even lower levels of atmospheric CO_2 (about 400 ppm).

Average atmospheric CO_2 levels have risen from 330 to 398 ppm over the last 35 years (Figure 9) and look set to pass through 400 ppm in the next few years and at the current rate of increase levels will be over 500 ppm in the next 50 years, and likely much sooner as global population increases along with growing demand for fossil fuels. Yet, we continue to take little action to plan for the impending changes to our environment let alone taking any action to mitigate the change.

When it comes to the big environmental questions facing the planet, so far it has been science and scientists that have begun the international collaborations necessary to face the challenges. In some ways, the most unnatural diplomats,

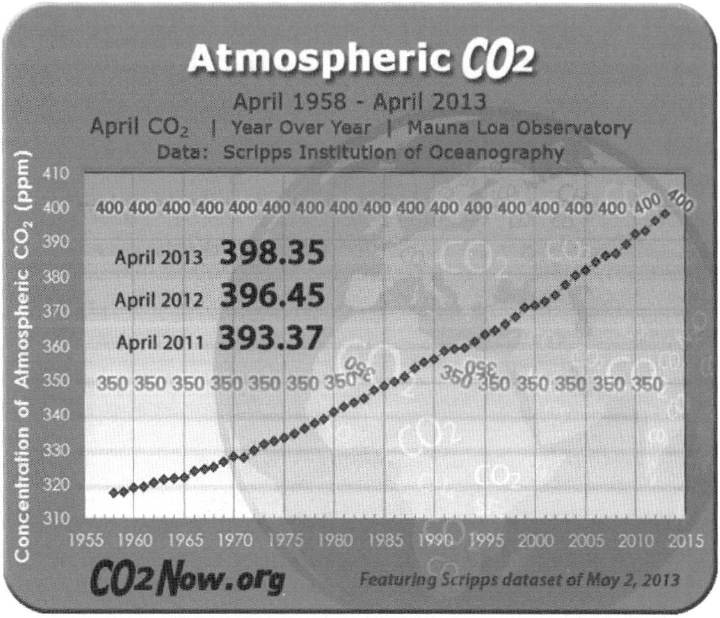

Figure 9. History of Atmospheric levels of CO_2 to July 2013 measured at the Mauna Loa Observatory.
Note: From http://www.co2now.org.

the scientists, become the deal brokers and political representatives have been able to capitalise on their initial international collaborations (science for diplomacy). The political negotiations have then provided the platform for strong multinational efforts to the benefit of all member countries (diplomacy for science).

With climate change, we are beginning the cycle again: the problem is global and beyond the scope of any single nation. An internationally collaborative approach is required to determine the extent of the problem and any steps for mitigation. The scientists have again led the way, with their results showing the need to take action as average atmospheric levels of CO_2 are set to pass 400 parts per million within the next two years. But our policy makers seem to be paralysed in their decision-making. Perhaps they cannot deal with the potential scale of global impact; perhaps they cannot enact policies that have a 50–100 year timeframe when they are influenced so strongly by much shorter electoral cycles? One thing is sure: the costs of living with the impacts of global warming will only increase if we wait. The lessons of IGY and ANDRILL are that it will just take one nation to take the lead and the pressures of diplomacy will ensure that others follow. Thereby, we can all reap the rewards that better diplomatic support for scientific endeavours can deliver.

References

Belanger, D.O. (2006) *Deep Freeze — The United States, the international Geophysical Year, and the Origins of Antarctica's Age of Science*, Boulder: University Press of Colorado.

Bentley, C.R., Cameron R.L., Bull C., Kohma K. and Gow A.J. (1964) 'Physical characteristics of the Antarctic ice sheet', *Antarctic Map Folio Series, 2*, New York: American Geographical Society.

Berkman, P.A. (2001) *Science into Policy*, San Diego: Academic Press.

Bonner, W.N. (1987) 'Antarctic Science and Conservation — The historical background', *Environment International*, 13: 19–25.

Broecker, W.S. (1991) 'The great ocean conveyor', *Oceanography*, 4(2): 79–89.

Cherry-Gerrard, A. (1922) *The Worst Journey in the World, Antarctica 1910–1913*, Facsimile Edition, London: Vintage Classics.

DeConto, R.M. and Pollard D. (2003) 'A coupled climate-ice sheet modeling approach to the Early Cenozoic history of the Antarctic ice sheet', *Palaeogeography, Palaeoclimatology, Palaeoecology*, 198 (1–2): 39–52. DOI: 10.1016/S0031-0182(03)00393-6.

Huxley, L. (1913) *Scott's Last Expedition (Volume 2): The reports of the journeys and the scientific work undertaken by Dr E.A. Wilson and the surviving members of the expedition*, New York: Dodd, Mead and Co.

Intergovernmental Panel on Climate (IPCC) (2013) *Climate Change. The Physical Science Basis: Contribution of Working Group 1st to the 5th Assessment Report of the Intergovernmental Panel on Climate Change*, Cambridge: Cambridge University Press.

Jones, A.G.E. (1981) 'The British Southern Whale and Seal Fisheries: A general overview', *The Great Circle*, 3(2): 20–29.

McKay, R., Browne G., Carter L., Cowan E., Dunbar G., Krissek L., Naish T., Powell R., Reed J., Talarico F. and Wilch T. (2010) 'The stratigraphic signature of the late Cenozoic Antarctic Ice Sheets in the Ross Embayment', *Geological Society of America Bulletin*, 121(11–12): 1537–1561.

Naish, T., Powell R., Levy R., Wilson G., Scherer R., Talarico F., Krissek L., Niessen F., Pompilio M., Wilson T., Carter L., DeConto R., Huybers P., McKay R., Pollard D., Ross J., Winter D., Barrett P., Browne G., Cody R., Cowan E., Crampton J.,

Dunbar G., Dunbar N., Florindo F., Gebhardt C., Graham I., Hannah M., Hansaraj D., Harwood D., Helling D., Henrys S., Hinnov L., Kuhn G., Kyle P., Läufer A., Maffioli P., Magens D., Mandernack K., McIntosh W., Millan C., Morin R., Ohneiser C., Paulsen T., Persico D., Raine I., Reed J., Riesselman C., Sagnotti L., Schmitt D., Sjunneskog C., Strong P., Taviani M., Vogel S., Wilch T. and Williams T. (2009) 'Obliquity-paced Pliocene West Antarctic Ice Sheet Oscillations', *Nature*, 458: 322–328. DOI: 10.1038/nature07867.

National Maritime Museum (2000). *South: The Race to the Pole*, London: Cambridge University Press.

Pollard, D. and DeConto R.M. (2009) 'Modelling West Antarctic ice sheet growth and collapse through the past five million years', *Nature*, 458: 329–332. DOI: 10.1038/nature07809.

Preston, D. (1999) *A First Rate Tragedy: Captain Scott's Antarctic Expeditions*, London: Constable and Company.

Rahmstorf, S. (2002) '*Ocean circulation and climate during the past 120,000 years*', *Nature*, 419: 207–214. DOI: 0.1038/nature07809.

Seki, O., Foster G.L., Schmidt D.N., Mackensen A., Kawamura K. and Pancost R.D. (2010) 'Alkenone and boron-based Pliocene pCO2 records', *Earth and Planetary Science Letters*, 292: 201–211. DOI: 10.1016/j.epsl.2010.01.037.

Summerhayes, C.P. (2008) 'International collaboration in Antarctica: The International Polar Years, the International Geophysical Year, and the Scientific Committee on Antarctic Research', *Polar Record*, 44(231): 321–334. DOI: 10.1017/S0032247408007468.

Tingey, R.J. (1983) 'Heroic age geology in Victoria Land, Antarctica', *Polar Record*, 21(134): 451–457. DOI: 10.1017/S0032247400021641.

Turner, J., Bindschadler R.A., Convey P., Di Prisco G., Fahrbach E., Gutt J., Hodgson D.A., Mayewski P.A. and Summerhayes C.P. (2009). *Climate Change and the Environment*, Cambridge: SCAR.

Walton, D.W.H. and Doake C.S.M. (1987) *Antarctic Science*, Cambridge: Cambridge University Press.

Wilson, G.S., Naish T.R., Powell R.D., Levy R.H., Crampton J.S. and McMurdo ANDRILL Science Implementation Committee (MASIC) (2012a) 'Late Neogene chronostratigraphy and depositional environments on the Antarctic Margin: New results from the ANDRILL McMurdo Ice Shelf Project', *Global and Planetary Change*, 96–97: 1–8. DOI: 10.1016/j.gloplacha.2012.05.007.

Wilson, G.S., Levy R.H., Naish T.R., Powell R.D., Florindo F., Ohneiser C., Sagnotti L., Winter D.M., Cody R., Henrys S., Ross J., Krissek L., Niessen F., Pompillio M., Scherer R., Alloway B.V., Barrett P.J., Brachfeld S., Browne G., Carter L., Cowan E., Crampton J., DeConto R.M., Dunbar G., Dunbar N., Dunbar R.,

Eynatten H., Gebhardt C., Giorgetti G., Graham I., Hannah M., Hansaraj D., Harwood D.M., Hinnov L., Jarrard R.D., Joseph L., Kominz M., Kuhn G., Kyle P., Läufer A., McIntosh W.C., McKay R., Maffioli P., Magens D., Millan C, Monien D., Morin R., Paulsen T., Persico D., Pollard D., Raine J.I., Riesselman C., Sandroni S., Schmitt D., Sjunneskog C, Strong C.P., Talarico F., Taviani M. and Villa G. (2012 b) 'Neogene tectonic and climatic evolution of the Western Ross Sea, Antarctica: Chronology of events from the AND-1B drill hole', *Global and Planetary Change*, 96–97: 189–203. DOI: 10.1016/j.gloplacha.2012.05.019.

CHAPTER 5

Diplomacy for Science: The SKA Project

Maria Pozza

Introduction

Science diplomacy may be defined as interstate scientific collaboration which aims at the constructive resolution of common problems in a co-operative manner. Science has historically played a fundamental role in establishing security and co-operation between states. One example was the announcement of the International Geophysical Year (IGY) by the scientific community in 1957 (which cooperative research extended thorough to 1958). Globally significant issues such as the high seas/oceanic environments, climate change and the outer space environment (to name but a few) offer much potential for a greater role for science and diplomacy in the future. However, continuing global tensions encompassing culture, religion, politics and economics have seen a dramatic change in post-Cold War diplomacy, especially in the fields of international relations. During the Cold War era, much energy was focused by states on research and development (R&D) for military purposes. The bipolar 'balance of terror' established by the superpowers emphasised military power and hard power politics, and under the JFK Administration generated the Cuban Missile Crisis. Such hard power considerations lingered through the 1980s. R&D objectives as advocated by powerful states during this time formed parts of their foreign policy, as was seen in the nuclear arms race and President Reagan's Strategic Defence Initiative or SDI — 'Star Wars'. An emphasis on military power tended to encourage the political application of scientific advances and R&D. In contrast, interstate scientific collaborations tend to encourage a soft power, or non-military, political approach. Post-Cold War diplomacy has tended to rely more upon collaborative efforts through science diplomacy utilising *smart power*. This is not intended to diminish the many highly-successful scientific collaborations that occurred during the Cold War. Such scientific collaborations have perhaps seen even greater success in the post-Cold War era however. With regard to science diplomacy, Flink and Schreiterer note that 'SD [science diplomacy] has become a proxy for both the use of science for diplomacy and of diplomacy for science, that is, the enhancement of scientific research and innovation capacities by way of international collaboration with mutual benefits' (Flink and Schreiterer, 2010: 666). In the post-Cold War era, both science and diplomacy rely upon each in a fluctuating equilibrium.

This chapter examines the Square Kilometre Array project (SKA) between Australia and South Africa as a case in point, in order to illustrate that science is not only a tool which might be used in diplomacy, but should be considered as a fundamental asset for improving international relations between states. SKA, which began in 1996, represents a good example of a collaborative scientific project being initiated and pursued by scientists rather than just states. The SKA project, which is being funded by 20 states, thus represents an evolution of science diplomacy. Through such science diplomacy and state cooperation led by specialist scientists, the SKA project aims to answer fundamental questions relating to the origin of the universe. The combined Australian-New Zealand efforts to host the SKA have seen further diplomatic and scientific cooperation, including the recent joining of the venture by former competitor South Africa. Through science diplomacy, the SKA project is now guaranteed to be one of the largest land-based efforts directed towards scientific questions concerning outer space. Science diplomacy will certainly continue to have a prominent role in the ongoing development of the project.

The Increasing Role of Diplomacy for Science

Chatterjee (2007: 80), commenting on the definition of *diplomacy*, notes that:

> ...diplomacy stands for the management of international relations, that is, primarily state-to-state relations. *'Management'* in this context would mean settlement of differences, which should be achieved by negotiation; the

> methods by which these relations are adjusted by ambassadors and envoys; the business or art of the diplomatist.[1]

Sharp (2009), writing from a political science standpoint, suggests that 'Diplomacy is conventionally understood to be the practice by which states represent themselves and their interests to one another'. From a historical and practical point of view, Hamilton and Langhorne (2011: 1) describe diplomacy as 'the peaceful conduct of relations amongst political entities, their principles and accredited agents...'.

Within the modern political global system, the United Nations allow state interactions which require diplomats to be in constant communication with home governments, and also to be up to date on all relevant issues. The economic interdependence of the global system warrants this requirement, and Chatterjee (2007: 81) notes that this is the root of bilateral diplomacy'. Diplomacy, as a dialogue between states, is a broad concept which ensures that the rapidly changing character of diplomacy within the international arena is catered for. The United Nations system allows diplomatic initiatives within its multilateral organs, and holds conferences through organs such as the International Telecommunication Union (ITU), the International Civil Aviation Organization (ICAO) and the International Atomic Energy Agency (IAEA). Diplomacy utilises such processes and negotiations concerning the representation of states interests, and applies soft power when necessary in order to find agreement on issues of concern.

Upon the subjects of diplomacy and its history, Berridge (2002: 1) writes that:

> Diplomacy is an essentially political activity and, well resourced and skilful, a major ingredient of power. Its chief purpose is to enable states to secure the objectives of their foreign policies without resort to force, propaganda, or law. It follows that diplomacy consists of communication between officials designed to promote foreign policy either by formal agreement or tacit adjustment. Although it also includes such discrete activities as gathering information, clarifying intentions, and engendering goodwill, it is thus not

[1] Continued from Chatterjee's definition which ended on the word 'negotiation'. It should be noted though that The Oxford English Dictionary entry as to the definition of diplomacy was also endorsed by senior British diplomat Sir Harold Nicolson.

> surprising that, until the label 'diplomacy' was affixed to all of these activities by the British parliamentarian Edmund Burke in 1796, it was known most commonly as 'negotiation' ... Diplomacy is not merely what professional diplomatic agents do. It is carried out by other officials and by private persons under the direction of officials. As we shall see, it is also carried out through many different channels besides the traditional resident mission. Together with the balance of power, which it both reflects and reinforces, diplomacy is the most important institution of our society of states.

Diplomacy, including the skills of negotiation and customary practices, rests on the core assumptions of sharing ideas and conversing on matters of importance to states, including those relating to the global system, state interests and the balancing of power. Secure communication and interstate representations are of core importance to diplomacy at the international level. International cooperation has grown since the 1950s, during which the United Nations has developed and procured a greater range of responsibilities over matters relating to the functioning of the international system, including matters pertaining to interstate relations. One example is the work of the International Monetary Fund (IMF), an institution that has been particularly active during recent decades. Much of the work within this body has concerned state representation, with diplomats entering into negotiations and utilising their diplomatic skills in order to address the economic concerns of states. For example, IMF has been active with European states, some of which have been economically stressed by the global recession.

Science Diplomacy

Assessing the meaning of diplomacy within the parameters of science requires some elaboration of the latter term. Diplomats usually have a good working knowledge of general concerns, but are not well equipped to understand the fine details of modern science required to understand scientific and technological developments. Science is defined by Copeland (2009: 112) as something which 'is predicated upon the empirical method of rigorous experimentation and the demonstrable repetition of results ... Science proceeds from the premise that all effects have a cause and, moreover, that these causes can be determined'. Scientific advice is crucial for diplomats and has seen the rise of scientific advisory bodies since the 1950s (National Research Council, 2002: 6). In 1957 and 1958 a global community of scientists joined together in a collaborative sharing of information and research under the auspices of the International Geophysical Year (IGY). The International Council of Scientific Unions arranged for considerable scientific collaboration across borders: Soviet scientists worked with American scientists on problems relating to outer space. Sixty seven states participated in the IGY by prior international agreements which were settled by the negotiation of diplomats. With the success of the IGY collaboration, other scientific research programmes arose which have led to institutions such as the Scientific Committee on Antarctic Research. Science is by its very nature a collaborative activity, and such collaborations may easily cross international boundary lines. The Cold War period involved cooperative scientific research with collaboration between the scientists of many states. This type of information sharing between scientists represents a good example of science diplomacy. Regardless of the Cold War context, scientists from both the Soviet Union and the US, collaborated and shared ideas concerning the exploration of outer space. Scientists have sometimes been instrumental during times of conflict in maintaining a bridge of cooperation and dialogue between states. Many Cold War era scientific groups established upon the basis of science diplomacy remain active in the post-Cold War era, such as the International Atomic Energy Agency (IAEA). Post-Cold War scientific diplomatic groups which have influenced states diplomatic relations with each other, and physically set international agendas for states to abide by, are exemplified by the 1997 Kyoto Protocol (extended to 2020 at the Doha Climate Change Conference).

The report on *New Frontiers in Science Diplomacy: Navigating the Changing Balance of Power,* published by the UK Royal Society and American Association for the Advancement of Science (AAAS), states that whilst science has had a role in the development of hard power politics, it 'primarily draws on "soft power"...: its attractiveness and influence both as a national asset, and as a universal activity that transcends national interests' (The Royal

Society, 2010: 11). The report states that 'science diplomacy' is a fluid concept which can usefully be applied to the role of science, technology and innovation in three dimensions of policy:

- informing foreign policy objectives with scientific advice (*science in diplomacy*);
- facilitating international scientific cooperation (*diplomacy for science*);
- using science cooperation to improve international relations between countries (*science for diplomacy*).

This chapter concentrates on *diplomacy for science*. It should be noted that a shift in paradigm has been advocated by Nye towards 'smart power' in relation to foreign policy and international relations, and that this shift has a direct impact on the application of this framework. Nye (2004: 5) works within the paradigm of a distinction between *hard power* (states utilising military and economic means to force or threaten states to act in a particular way), and *soft power* (which builds upon states' common interests and values in order to find common ground and persuade or influence). He promotes smart power as a blending of hard and soft power thus: 'Soft power is the ability to get what you want through attraction rather than coercion or payments of hard power. Knowing how to combine hard and soft power instruments is smart power' (Guopeng, 2009). *Diplomacy for science* sits well within such a framework and has arguably been operating for some time. Science can confer hard power capability upon states, as was seen with the development of the atom bombs used to destroy Hiroshima and Nagasaki in 1945, contributing to the Japanese surrender of World War II.[2] However, in ways unlike those of traditional international relations theory and traditional diplomatic methods, science also supplements hard power with soft power potentialities. Examples include organisations such as Scientists against the Nuclear Bomb, Union of Concerned Scientists, and large collaborative

[2] A further illustration of hard power (state control of scientists) may be seen post-WWII when scientists from Germany were accepted into the service of US and former USSR. Both the US and former USSR-employed German scientists who had previously worked for the Nazi rocket programme and installed them in their own rocket programmes in post war years (US Project Paperclip).

efforts by increasing non-state science actors such as the Radio Frequency Council and the SKA organisation project committee. The latter two examples illustrate how such organisations have been operating smart power policies in order to pursue cross border collaborations and work on high risk scientific projects in order to advance science.

Diplomacy for science as practised by states, inter-state departments, non-governmental organisations and scientists, and including the smart power approach, is increasing. As Leguey-Feilleux (2009: 14) notes, 'non-governmental organizations (NGOs) are components of civil society ... some NGOs do not limit their endeavours to domestic societies. They network across international boundaries...'. These trans-national bodies, which include science organisations, work within the structure of the state but are free to cross borders.[3] Non-state science actors are able to utilise the principles of smart power by fostering cross boundary relations and partnerships in order to develop inter-state scientific applications or technological innovations within the realm of science diplomacy.

SKA: An Example of Diplomacy for Science

Diplomacy for science performs a strong role within the international arena, which has been developed over time. Flink and Schreiterer (2010: 665) note that 'scientific cooperation comes to be seen as an effective agent to manage conflicts, improve global understanding, lay grounds for mutual respect and contribute to capacity-building in deprived world regions'. This consideration is applicable to South Africa's involvement in the SKA project.

SKA is a radio telescope comprising an array of antennas strategically placed in open areas free from '*radio noise*', in order to relay information to two separate central cores, located in Australia and in South Africa. The Australian element of SKA is planned to cover over 3,000 kilometres of mid-western Australia with the core to be established at the Murchison Radio Astronomy Observatory. The SKA project aims to cover 5 key scientific programmes: Probing the Dark Ages, Galaxy Evolution, Cosmology and Dark Energy; the Origin and Evolution of Cosmic Magnetism, Strong Field Tests of Gravity Using Pulsars and Black Holes; Probing the Cosmic

[3] Note also growing role of NGOs in international security; see Aall (1996).

Dawn; and the Cradle of Life (Carilli and Rawlings, 2004). The SKA project represents a further advance on the present Australian Square Kilometre Array Pathfinder (ASKAP) project. The Commonwealth Scientific and Industrial Research Organisation (CSIRO) will direct and implement the SKA project as is the case with ASKAP.

The SKA project is a prime example of science diplomacy involving many nations with strong scientific links to each other. The genesis of the SKA project lay with the scientific communities' call for a large hydrogen telescope. The International Union of Radio Science (URSI) established a 1993 working group whose objectives were to study next-generation radio wavelength emissions, and aimed at the construction of an observatory to this end.

Dewdney *et al.* (2009) wrote that:

> Advances in astronomy over the past decades have brought the international community to the verge of charting a complete history of the Universe. In order to achieve this goal, the world community is pooling resources and expertise to design and construct powerful observatories that will probe the entire electromagnetic spectrum, from radio to gamma-rays, and even beyond the electromagnetic spectrum, studying gravitational waves, cosmic rays, and neutrinos.
>
> The Square Kilometre Array (SKA) will be one of these telescopes, a radio telescope with an aperture of up to a million square meters. The SKA was formulated from the very beginning as an international, astronomer-led (grass roots) initiative. The International Union for Radio Science (URSI) established a working group in 1993 to study the next-generation radio wavelength observatory. Since that time, the effort has grown to comprise 19 countries and 55 institutes.

The project was eventually decided to be best suited to one of two regions on the planet: Karoo, Central South Africa, or Western Australia. New Zealand's south Island was ruled out of consideration in 2012 as a site, but New Zealand remains a partner with the Australian venture. In 2012, the international SKA Site Advisory Committee decided that both Australia and South Africa would be host sites with separate core's acting as central analysis hubs. Headquarters are situated in Manchester, UK. With a collaboration of 20 nations, two core sites and head office located outside of the core facilities,

Hall *et al.* (2008: 4), correctly notes that 'the SKA was "born global"'. The estimated cost of 1.5 billion euros will be shared by the countries concerned, and the project 'is aimed to provide answers to fundamental questions about the origin and evolution of the Universe' (Redfern, 2011). As well as the many states involved in the project, private and public corporations as well as governmental bodies are also engaged with a community of 55 institutes and approximately 200 scientists and engineers (Hall *et al.*, 2008).

This chapter can only endeavour to describe the SKA project as administered and managed between Australia (Department of Innovation, Industry, Science and Research and CSIRO) and South Africa (Department of Science and Technology (DST), and SKA Africa), and will not examine the role of non-state actors. It is significant that the CSIRO is the Australian national science agency. As the national government body for scientific research in Australia, the CSIRO exhibits science diplomacy in its interactions with other organisations and states. Similarly the Department of Science and Technology, funded by the National Research Foundation, will direct and implement the SKA project in South Africa. It should be noted that both the CSIRO and the DST are functioning in regard to SKA as inter-governmental scientific agencies and are thus not entirely separated from state interests within the global arena. The SKA project will work within the parameters of Australia's and South Africa's interests, as well as increasing the strength of international relations between the two nations, as well as the other twenty nations which have been involved with the SKA project.

Below is a discussion of some of the challenges both the Australian CSIRO and South African DST may face nationally, as well as an overview as to how the paradigm of diplomacy for science will affect these challenges.

Australia: SKA and the Role of Diplomacy for Science

SKA is a good illustration of diplomacy for science as a means for the Australian government to facilitate international scientific cooperation, which can then result in improved international relations. Initially Australia entered the SKA bid with New Zealand. It was hoped that New Zealand would be used as one of the sites for the radio telescope. In light of the later South African bid, Australia (with New Zealand support) shifted in its approach and began discussions with SKA South Africa as to the possibility of a joint proposal. This eventually led to the joint Australian-South African SKA submission. This approach was certainly in line with the Australian *Science and Industry Research Act 1949* (henceforth SIRA 49). There are two important sections under this piece of legislation in relation to science diplomacy and the Australian effort on the SKA project:

1) section 9(1)(c) stipulates that 'the functions of the Organisation are to act as a means of liaison between Australia and other countries in matters connected with scientific research';
2) section 10 further stipulates that 'the Organisation shall, as far as possible, co-operate with other organizations and authorities in the co-ordination of scientific research, with a view to:
   (a) the prevention of unnecessary overlapping'[4]

Overall, SIRA 49 adopts and embodies many of the principles of science diplomacy within the legislation. It actively encourages CSIRO to engage in international cooperation on scientific matters and promote international dialogue/sharing of scientific information. However, CISRO has a primary obligation to Australia and Australian interests. Section 9(1)(a)(i–iv) stipulates that scientific research is for the purposes of the Australian community and Australian national objectives or its performance concerning responsibilities to the Commonwealth or any other purposes as determined by the Minister. The joint SKA venture with South Africa is thus reflective of the Australian objective to optimise its scientific investments into scientific research and infrastructure, which will also aid national security (Australian Academy of Science, 2011: 15). It is important to note that whilst CISRO is committed to science diplomacy as a borderless conversation, SKA is still representative of Australian national interests and that this is an issue which may affect future decisions concerning sharing and dissemination of information attained through the SKA.[5]

---

[4] Australia. Attorney-General's Department (SIRA 49) (2012) with Act No. 13 of 1949 as amended 28th March 2012, including amendments No.89 of 2011).

[5] Section 9(1)(h–i) stipulates that CSIRO's purposes is to *disseminate information relating to scientific matters and publish journals*, however section 9(2)(a–b) is clear in that CSIRO's must treat Australia's interests and objectives as its primary function.

Thus due to CSIRO's governmental and national interests, the dissemination and sharing of information may theoretically become an issue due to potential conflict between international (other state) and national concerns. Whilst the other twenty states financially contributing, to the project will have national interests as well, the CSIRO is the lead scientific organisation within the group and thus has greater determination power. That power will of course be influenced by its national interests. This is illustrative of the parameters in which the SKA in its diplomacy for science approach, will see limits within which it must operate. Australia is nevertheless in a unique position in relation to international scientific collaboration which can lead to a strengthening of international relations.

Similarly, SKA has utilised tools from diplomacy for science in relation to the Aboriginal community and the land to be used for the SKA project. Although the Australian site bid dealt with the complexities surrounding land title on the proposed site, Peter Viney, former Australian SKA committee chairman until 2009, stated to the South African media that this continued to be a potential concern. Viney noted that there is not yet a native title agreement to use the land protected for Aboriginal populations in relation to the Western Australian site at Murchison. 'Other native title negotiations for major resource projects in Western Australia, such as the acquisition of land in the Kimberley area for a natural-gas processing plant, have been very difficult, have involved significant protests and controversy between affected Aboriginal people, and have involved costs exceeding AUS$1billion' (Carpenter, 2012). The Wajarri Yamatji Native Title Claim (WY Claim) includes land that the ASKAP and future SKA project will require to be developed.

The Murchison Radio-astronomy Observatory Agreement was signed on 13 November 2009. CSIRO thus entered into an agreement with the Yamatki Marlpa Aboriginal Corporation (YMAC) for the use of this land. This agreement names CSIRO as the Crown Lease holder for the purposes of both the ASKAP and the SKA projects (known as the projects) and falls under the category of an Indigenous Land Use Agreement. The agreement was necessary given that the projects extend to land covered by the WY Claim. The agreement further holds that 'the grant of this lease does not extinguish the native title rights and interests of the native title claimants, however, such rights and interests are suspended to the extent that these are inconsistent with the rights conferred by the lease' (Murchison Radio-astronomy

Observatory Agreement, 2009). The extent of the suspension of such rights under Clause 4.4 'Non-Extinguishment Principle' (Murchison Radio-astronomy Observatory Agreement, n.d.: 17) is however not clear. Similarly under Clause 4 'the consents to future acts' also remain ambiguous. The ambiguity within the agreement allows flexibility of interpretation, especially in relation to future circumstances; however, this could cause problems for either party to the agreement. Without clear definitions, deviation from the agreement by either party could be difficult to determine and this could create further problems, especially in relation to future development of the land in question.

Supplying power to the remote areas in which the SKA projects are located in Australia and South Africa will also be challenging. The energy demanded by the projects means that diplomatic industrial negotiations are underway promoting the scientific benefits of the project. The design goals of the SKA project will need to take these difficulties into consideration. If energy supply is not readily available to the sites, then the SKA projects may be compromised. The cost of power supply will also need to be considered as another potential limiting factor for the projects. The MeerKAT array in South Africa and the ASKAP array in Australia have provided excellent testing grounds for the potential energy consumption necessary to fuel the projects. Evidence from these sites has provided forecast information for the Power Investigation Task Force (PITF), now disbanded, which evaluated the energy challenges (SKA Project, n.d.). The PITF 'was formed to raise awareness of the issue and, in particular, to forge links between the electronics-dominated SKA engineering community and strategic thinkers in the power industry'.[6] Considerations undertaken concerned both alternative power sources and which elements of the SKA projects would require continuous and uninterrupted power, in contrast to those instruments where the high integrity of power supply may not be fundamental. For example, the cores of the SKAs will need high integrity and uninterrupted power flow, whilst the antennas may require less continuous power supply. These problems will potentially impact the design features of the SKA project at the sites.

---

[6] Information concerning this disbanded body was extremely limited on the www.skatelescope.org website including findings by the committee. Hall P. J. (2011: 1) provided some excellent insights into the work of this group and his work is highly recommended.

The *diplomacy for science* framework will be crucial here, especially in relation to the care and skill necessary when undertaking assessments regarding which power supply or multiple power supplies would be more advantageous for the projects. Design features affected as a result of power supply may necessitate negotiations in relation to other technologies or scientific methods which might be adopted by private power companies, or in relation to imported technologies from other states. Such developments will also need to bring to the attention of the supporting countries the costs involved in development of technologies, as well as potential benefits which may result. There is a very real possibility that non-state actors or private energy corporations may need to have a recognised role within the SKA project and become part of the process and development of the SKA. The development of an effective power supply may also see the need for effective international negotiations. The CSIRO may need to advocate and negotiate between the energy industry and international project holders as to the benefits of such industry development. Hall (2011: 1) makes the sound point that 'provision of reliable power is essential for scientific effectiveness of a radio telescope ... With construction of ASKAP [and] MeerKAT ... the SKA community is learning from efforts at providing affordable, reliable power at much more remote and inhospitable sites'.

Values such as transparency are integral to the diplomacy for science framework as a means to building stronger international relations. The above issues will test this value, but it should be noted that SKA is well positioned to build frameworks of trust and cooperation through SKA which may translate into strengthened international relations.

South Africa SKA: The Growth of Diplomacy for Science

The South African bid to serve as a host site for the SKA was shortlisted in 2006, alongside that of Australia and New Zealand. South Africa currently has the MeerKAT array in operation in the Karoo region. The MeerKAT project is a basis for future development of the SKA. Whilst the Australian site will monitor low-frequency radio waves because of the advantageous 'radio quiet' characteristics in the surrounding area, South Africa offers the project a suitable location from which to monitor the medium and high frequency radio waves due to the nature of its geographical location. South Africa's nomination as a SKA project site in part recognises the increasing political stability of this country, as well as representing international recognition of South African efforts in regards to science and radio astronomy.

In 1996, a white paper (White Paper on South Africa, 1996) was released by South Africa's first democratic government which recognised the importance of the role science diplomacy would play for South Africa's continued growth. A new South Africa is growing, utilising the tools of science diplomacy to do so, and Pandor explains that 'one of the current flagship areas for South African science diplomacy, namely radio astronomy... is responsible for providing Africa with a substantial new cohort of scientists, engineers, technicians, and other knowledge workers' (Pandor, 2012). Similarly, in relation to space, South Africa's National Space Agency places the country as an emerging space faring state especially in relation to its aims of utilisation of space for observation, communication and navigation purposes. The creation of the Department of Science and Technology (DST) in 2002 is a further illustration of the weight attached by the South African state to the importance of science diplomacy. One of the aims of the department is to promote scientific cooperation; it stipulates that it 'aims to develop, promote and manage strategic international relationships, opportunities and science and technology agreements that strengthen the National System of Innovation and enable an exchange of knowledge, capacity and resources between South Africa and its regional and other international partners' (ZADST, n.d.).[7]

However, future investment into other state assets such as natural gas may become a factor which might interfere with the SKA project. For example, in 2010 Royal Dutch Shell made an application to explore 90,000 km^2 in Karoo region in their search for natural gas reserves. Val Munsami raised questions during the parliamentary session concerning Shell's plans for development of natural gas reserves in South Africa. Munsami stressed that legislation is in place to protect and regulate radio waves and prevent interference. Munsami further noted that 'obviously we will be looking at whether, in terms of exploration, there is any radio interference. If there is, we will have to have that discussion in terms of the regulatory framework' (TechCentral, 2011). South Africa has implemented the Astronomy Geographic Advantage

[7] Also see South Africa Corporate Strategy (2009/10).

Act of 2007 and thereafter the regulations to prohibit or restrict certain activities in core astronomy advantage areas in terms of the Act (Astronomy Geographic Advantage Act, 2007). It may face problems if there is significant flexibility in relation to exploratory rights of multinational corporations. The present South African legislation in place is to protect specific radio waves designated for radio astronomy, and legislates against interference with radio waves, but does not specifically prohibit other activities such as land development in the region which might pose a future threat to the SKA project.

The role of *diplomacy for science* in South Africa is not as well established in comparison to Australia, but is growing. The 2011 South African White Paper on Foreign Policy stipulates that South Africa's international engagement in international relations is built upon cooperation, multiculturalism and dialogue through strengthening already existing bilateral relations (White Paper on South Africa, 2011: 5). Since the establishment of democratic South Africa in 1994, it has focused much of its energy on Afro-Centric Foreign policy which has seen a rise in the importance of science diplomacy, especially *science for diplomacy* (White Paper on South Africa, 1996). Science and technology has been a major focus for South Africa for both economic growth and competitiveness, and developing science partnerships has been a priority for this state (White Paper on South Africa, 1996). During her time as Minister of Science and Technology, Pandor (2012) stipulated that, 'The growing importance of the science content of critical foreign policy issues has necessitated that the South African government pursue a concentrated science diplomacy strategy'.

The SKA project provides an excellent opportunity for South Africa to demonstrate its growing expertise in both science and science diplomacy. Pandor (2012) noted in relation to the SKA project that 'this is no small achievement for the science diplomacy efforts of South Africa and its partners, to have a discipline, traditionally viewed as an elite basic research domain dominated by developed countries, now being recognized at the highest level as a flagship initiative not only for African scientific capacity building, but also for broader regional integration and economic development'.

Whilst South Africa may still be viewed as an emerging scientific state, it has utilised many of the tools of *diplomacy for science* in order to build strong international relations based on science through its DST which includes membership to the Brazil, Russian Federation, India, Chinese and South

Africa forum (BRICS). BRICS is a platform upon which South Africa will build stronger partnerships through international relations and science diplomacy. South Africa's active involvement with the world Summit on Sustainable Development, as well as with the International Council for Science will also see South Africa utilise *diplomacy for science* tools.

Australia and South African Diplomacy for Science Relations

Australia and Africa have enjoyed strong bilateral relationships, including research links, and the SKA project promises growing recognition of diplomacy for science in the region. Australian research institutions and governmental agencies have had much involvement with African research and development. Since 2012 ACIAR in partnership with Canada's IDRS have made a commitment towards the cultivation of Africa's future through science, in a good example of agricultural science diplomacy. The Australian Official Development Assistance (ODA) has thus assisted agricultural science in Africa, in-conjunction with the Australian Centre for International Agricultural Research (ACIAR), and the Australian Nuclear Science and Technology Organisation (ANSTO) has worked closely with South Africa. Like CSIRO, ACIAR is a statutory body whose primary goal is to advance Australian interests. However, the utilisation of diplomacy for science and collaborative efforts have seen strong international relations develop between Africa (especially South Africa) and Australia. The SKA provides a unique opportunity for further collaboration and a basis upon which mutually beneficial programmes might be established. Australian and South African SKA proposed sites offer unique insights into the SKA scientific research to be conducted. The sharing of information between these states will only strengthen international relations between the two nations and will inevitably flow onto the other collaborating SKA states.

Utilising the tools of *diplomacy for science,* stronger relations will be created between the developed state of Australia and developing state of South Africa. Through the utilisation of science and science diplomacy, the SKA project will aim to answer major scientific questions as well as improving international relations. South Africa, as a newly democratic nation, is still emerging as a science power, and is attaining recognition in an area of science which is often dominated by other nations. Australia will benefit from the collaboration also. The two sites of the joint SKA project offer the promise of greater scientific findings. Although both states face potential problems in relation to the implementation of the SKA, the joint venture illustrates the potential of science diplomacy to further international relations and cross geographic as well as scientific boundaries.

Conclusion

Science diplomacy is a dynamic new paradigm of collaboration and development that is highlighted by the SKA project under development in Australia and South Africa, with headquarters in the UK. The SKA project is expanding the international links between the countries in question, facilitating scientific cooperation and informing foreign policy objectives for the countries concerned. Science diplomacy is growing in importance due to modern scientific endeavours such as SKA, and is gaining official recognition from states as a valuable tool in the development of science and international relations. It encourages communication and in so doing breaks down barriers, introducing new layers of interaction for international relations. There are some limitations to science diplomacy but it does provide a sound basis for development and a positive and constructive model for international relations in the future through a framework of *diplomacy for science*.

Cited References

Aall, P. (1996) 'Nongovernmental Organization and Peacemaking', in Crocker C.A., Hampson F.O. and Aall P. (eds). *Managing Global Chaos: Sources of, and Responses to, International Conflicts*, Washington, DC: Institute of Peace, pp. 433–444.

Astronomy Geographic Advantage Act (2007) *Astronomy Geographic Advantage Act 21 of 2007*, [Online], Available: http://www.saflii.org/za/legis/consol_reg/agaa21o2007rangnr465740.pdf.

Australia. Attorney-General's Department (SIRA 49) (2012) *Science and Industry Research Act of 1949*, Canberra, Australia: Office of Legislative Drafting and Publishing.

Australian Academy of Science (2011) 'Australian Science in a Changing World: Innovation Requires Global Engagement', Position paper, November 2011, Canberra, ACT, Australia, Available: http://www.science.org.au/sites/default/files/user-content/innovationrequiresglobalengagement_2.pdf.

Berridge. G.R. (2002) *Diplomacy: Theory and Practice*, 2nd edition, Chippenham, Wiltshire, UK: Antony Rowe Ltd.

Carilli, C. and Rawlings S. (eds.) (2004) 'Science with the SKA', *New Astronomy Reviews*, 48(11–12): 979–1606.

Carpenter, A. (2012) 'Oz Telescope Body Under Microscope After Ex-Chairman Raises Difficult Questions', *The Star*, [Online], Available: http://www.iol.co.za/the-star/oz-telescope-body-under-microscope-after-ex-chairman-raises-difficult-questions-1.1239657#.UJYGvMXMgn8 [22 February 2012].

Chatterjee, C. (2007) *International Law and Diplomacy*, London: Routledge.

National Research Council (2002) *Knowledge and Diplomacy: Science Advice in the United Nations System*, Washington, DC: The National Academic Press.

Dewdney, P.E., Hall P.J., Schilizzi R.T., Joseph T. and Lazio, L.W. (2009) 'The Square Kilometre Array', Invited Paper, *Proceedings of the IEEE*, 97(8): 1482–1496.

Flink.T and Schreiterer U. (2010) 'Science Diplomacy at the Intersection of S&T Policies and Foreign Affairs: Toward a Typology of National Approaches', *Science and Public Policy*, 37: 665–677.

Gluckman, P. (2011) 'New Zealand Science and our International Connections; Science, "Globalisation" is the Future', *Office of the Prime Minister's Science Advisory Committee*, 22 February 2011, [Online], Available: http://www.pmcsa.org.nz/wp-content/uploads/Sir-Peter-Gluckman-speech-NZ-Greenhouse-Gas-Research-Centre-22-Feb-2011.pdf.

Guopeng, J. (2009) 'Harvard professor Joseph Nye: US must combine soft, hard power into "smart power"', *China View*, 16 January 2009, [Online], Available: http://news.xinhuanet.com/english/2009-01/16/content_10669964.htm.

Hall. P. J. (2011) 'Power considerations for the Square Kilometre Array (SKA) Radio Telescope', *General Assembly and Scientific Symposium, 2011 URSI Proceedings*, [Online], Available: http://www.ursi.org/proceedings/procGA11/ursi/J03-1.pdf, pp. 1–4. DOI: 10.1109/URSIGASS.2011.6051200.

Hall, P.J., Schilissi R.T, Dewdney P.E.F. and Lazio J.W. (2008) 'The Square Kilometer Array (SKA) Radio Telescope: Progress and technical directions', *International Union of Radio Science*, 236: 4–19. Available: http://espace.library.curtin.edu.au/webclient/StreamGate?folder_id=0&dvs=1379950325845~928&usePid1=true&usePid2=true.

Hamilton K. and Langhorne R. (2011) *The Practice of Diplomacy: Its Evolution, Theory and Administration*, 2nd edition, London: Routledge.

Murchison Radio-astronomy Observatory Agreement (n.d.) *Murchison Radio-astronomy Observatory Agreement: Indigenous Land Use*, [Online], Available: www.atns.net.au/objects/TGTXZCVZENV/Complete%20ILUA.pdf.

Nye, J. (2004) *Soft Power: The Means to Success in World Politics*, New York: Public Affairs.

Pandor, N. (2012) 'South African Science Diplomacy: Fostering Global Partnerships and Advancing the African Agenda', *Science and Diplomacy*, 9 March, [Online], Available: http://www.sciencediplomacy.org/perspective/2012/south-african-science-diplomacy [10 November 2012].

Redfern, M. (2011) 'World's biggest radio telescope, Square Kilometre Array', *BBC News*, 31 March, [Online], Available: http://www.bbc.com/news/science-environment-12891215 [10 November 2012].

The Royal Society (2010) *New Frontiers in Science Diplomacy: Navigating the changing balance of power*, January, [Online], Available: https://royalsociety.org/~/media/Royal_Society_Content/policy/publications/2010/4294969468.pdf

Sharp, P. (2009) *Diplomatic Theory of International Relations*, Cambridge: Cambridge University Press.

SKA Project (n.d.) 'Power Investigation Task Force (PITF) (Disbanded)', [Online], Available: http://www.skatelescope.org/the-project/history-of-the-organisation/committees-working-groups-2/power-investigation-task-force/ [11 November 2012].

White Paper on South Africa (1996) 'Science and Technology: Preparing for the 21st Century', *White Paper on Science and Technology*, 4 September, [Online], Available: http://www.esastap.org.za/download/st_whitepaper_sep1996.pdf [11 November 2012].

White Paper on South Africa (2011) Building a Better World: The Diplomacy of Ubuntu', *White Paper on South Africa's Foreign Policy*, Final Draft, 13 May, [Online], Available at: http://db3sqepoi5n3s.cloudfront.net/files/docs/110513SApolicy-foreign.pdf [10 November 2012].

Department of Science and Technology. Republic of South Africa (ZADST) (n.d.) *International Cooperation and Resources*, [Online], Available: http://www.dst.gov.za/index.php/internatprog [10 November 2012].

South Africa Corporate Strategy (2009/10) *Department of Science & Technology on Strategic Plan & Budget 2009/10*, 8 June 2009, Available: http://www.pmg.org.za/report/20090609-department-science-and-technology-strategic-plan-and-budget-vote-31 [10 November 2012].

TechCentral (2011) 'SKA concerns Over Shell's Karoo Gas Plans', *TechCentral Editorial*, 16 March, [Online], Available: http://www.techcentral.co.za/ska-concerns-over-shells-karoo-gas-plans/21911/ [10 November 2012].

Additional References

Alden, C. and Vieira M.A. (2005) 'The New Diplomacy of the South: South Africa, Brazil, India and Trilateralism', *Third World Quarterly*, 26(7): 1077–1095.

Commonwealth Scientific and Industrial Research Organisation (CSIRO) (n.d.) *CSIRO Operational Plan 2012–13*, [Online], Available: http://www.csiro.au/Portals/About-CSIRO/How-we-work/Budget--Performance/Operational-plan.aspx.

Hamill, J. and Lee D. (2001) 'A Middle Power Paradox? South African diplomacy in the post-apartheid era', *International Relations*, 15(4): 33–59 DOI: 10.1177/004711701015004004.

Leguey-Feilleux, J.-R. (2009) *The Dynamics of Diplomacy*, Boulder, CO: Lynne Reinner Publishers.

SKA Organisation (2012) *Members' Statement on Ratification of Site Agreement*, 14 November 2012, Manchester, UK. See www.skatelescope.org/thelocation/ and generally www.skatelescope.org.

Vale, P. (2012) 'Revealing All? The Troubled Time of South Africa's Diplomacy', *The Hague Journal of Diplomacy*, 7(3): 337–349. DOI: 10.1163/187119112X642953.

Wilson, K. (2013) 'The Curious Case of CSIRO's GM Field-Pea', *New Mathilda*, 1 May 2013, [Online], Available: https://newmatilda.com/2013/05/01/curious-case-csiros-gm-field-pea [20 May 2014].

Wolfe, A. J. (2013) 'Science Diplomacy Works, but only when it's Genuine', *The Guardian*, 23 August 2013, [Online], Available: http://www.theguardian.com/science/political-science/2013/aug/23/obama-science-foreign-policy [20 May 2014].

Part II

Diplomacy for Science: Informing Foreign Policy Objectives with Scientific Advice

CHAPTER 6

Science and Climate Change Diplomacy: Cognitive Limits and the Need to Reinvent Science Communication

Manjana Milkoreit

In a time of multiplying international problems that require scientific input, a well-functioning science-diplomacy interface is vital for the success of global governance. The case of climate change offers valuable lessons concerning current institutional design patterns of this interface, building on more than two decades of experience with of the Intergovernmental Panel on Climate Change (IPCC) and the United Nations Framework Convention on Climate Change (UNFCCC). In this chapter I depart from the usual approach to analysing the role of science in diplomacy. Instead of assessing the functioning and processes of the IPCC, I use a cognitive perspective to analyse how diplomats' minds receive and make use of scientific knowledge. Using interview data from 2012, I argue that: (1) most negotiation participants use a very basic and limited set of insights about climate change that has not changed significantly for a long time; (2) that recent scientific concepts — most notably the idea of climatic tipping points — are not yet part of most diplomats' belief systems; and (3) that hardly any negotiator is able to imagine qualitatively different long-term futures that have been affected by climate change, and to link present decisions to those possible futures. I discuss the implications of these findings for the negotiation process and outline possible ways to improve the design of the science-diplomacy interface to address present cognitive limitations.

Humans have barely begun to understand what living in the Anthropocene (Crutzen 2006; Steffen *et al.*, 2007) means. A key feature of this human-dominated epoch is the multiplication of governance challenges with extremely long problem time scales that create links between human action in the present and their effects in the distant future. Understanding the power of the present generations to influence the wellbeing of many generations to come is an unprecedented challenge for humanity; yet, it is a fundamental requirement for making 'good' decisions, especially in global governance. Rising to the challenge of good decision-making in the Anthropocene requires a set of cognitive skills that are currently woefully underdeveloped even in the brains of some of the brightest scientists and the most passionate diplomats: complex systems thinking, in particular understanding the dynamics of linked social-ecological systems, and imagining the distant future.

The increasing need of science in diplomacy is an inevitable side effect of the Anthropocene. But being in need of science refers not only to the availability of large amounts of high quality scientific information, but also to the ability to use this information in negotiation and decision-making processes. As I will argue in this chapter, this aspect of the relationship between science and diplomacy — uptake, understanding and use of scientific information by diplomats — is a larger problem than the provision of knowledge. Curiously, this 'receiving' side of the science-policy interface has not received much attention in the past, either by scholars or practitioners.

Climate change is a prime example for the type of diplomatic challenge on the rise and therefore an important test-bed for the functioning of the science-diplomacy interface. If one assumes that understanding the science of climate change is a necessary condition for the effective governance of climate change, what kind of science, both in terms of substance and communication, do we need?

This is where the IPCC comes into the picture. The IPCC's primary task is to provide the scientific information necessary for climate change diplomacy without making policy prescriptions. Much of the literature on the role of science in climate governance focuses on the IPCC (Hulme and Mahony 2010; Vasileiadou *et al.*, 2011), its functioning (Ho-Lem *et al.*, 2011), its successes and failures as a boundary organization (Hoppe *et al.*, 2013) and associated debates about the role of scientists between truth-to-power speakers and political advocates (Gamson 1999; Lackey 2007; Arimoto and Sato 2012).

The jury is still out on the IPCC's effectiveness. Their successive synthesis reports have certainly raised the profile of the climate issue since the late

1980s and accompanied a tremendous increase in efforts to create a global climate regime. Yet, so far the science has not been able to move the politics to a place where one can start measuring environmental effectiveness. One could argue that this is not the task of science — after all, the job of scientists is to provide information, and the job of diplomats and politicians is to make the decisions — but in the end, effectiveness is the ultimate measure the success of science in diplomacy.

In this chapter I will not talk about the IPCC and how it provides scientific information. Instead I tackle the complementary question of how science is received and used in the minds and beliefs of diplomats. Understanding this receiving end of science communication requires a very different theoretical approach than the institutional and process analysis linked to the IPCC. I use a cognitive approach to identify existing belief systems among participants in the UNFCCC negotiations, in particular their use and ignorance of scientific concepts and ideas.

Using findings from empirical work conducted in 2012, I argue that diplomats use only a very limited and simplified set of data points or facts about climate change in their daily work. Two observations are particularly noteworthy. First, there is a general lack of attention to the concept of climatic tipping points and its implications for climate governance. Second, the majority of study participants had a very limited ability to imagine possible futures and therefore lacked important motivational elements for making decisions on climate governance. Addressing each of these issues in more detail, I will conclude with some thoughts about the implications of these observations for the future of the science-diplomacy interface, and the need to reinvent science communication based on our growing understanding of the working of mind and the rules of cognitive change.

Theorizing about the Science-Diplomacy Interface

Research on the science-diplomacy relationship is embedded in larger debates about the relationship between science on the one hand and politics and policy-making on the other. In modern societies scientific knowledge is considered a necessary condition for enlightened decision-making: 'we presume that knowledge will lead to rational action' (Norgaard, 2009: 12). Making a distinction between science as the pursuit of truth and politics as a struggle between values, many scholars, policy-makers and voters alike assume that there is a rational or logical link between a scientific finding, such as anthropogenic climate change, and the 'right' policy response. Hulme calls this model of a science-policy interface 'technocracy' (Hulme, 2009: 102–103).

Over the last three decades, the debates both in- and outside the academy have shifted significantly away from this neat model of two separate spheres with a unidirectional information flow from science to politics, and the rational use of such information by decision-makers. A more complicated view has emerged that recognizes science as a social activity, which is shaped by the values, social habits and incentives of individual scientists (Parker *et al.*, 2010; Parker and Hackett, 2012), and also emphasizes the increasingly blurred boundaries and messy relationships between science and politics in various models of organizing scientific expert advice (Jasanoff, 2004; Lackey, 2007; Miller, 2007). This debate has received impetus from scholarship on wicked problems (Rittel and Webber 1973; Verweij *et al.*, 2006; Levin *et al.*, 2012), post-normal science (Funtowicz and Ravetz, 1993) and resilience (Folke, 2006), but also by political debates about planetary boundaries (Pielke, 2013; Galaz, 2013) and, most importantly, the role of climate change scientists between research and advocacy (Lackey, 2007). In the emerging picture the two spheres are linked in multiple, complex ways, which are often not very well understood, and a badly organized science-politics interface can have tremendous disadvantages for society.

As mentioned above, the IPCC acts as the key source for scientific information in the UNFCCC process. In a fundamental way, the IPCC's design and mandate reflect the technocratic model outlined below: the scientific body aggregates and synthesizes existing knowledge on climate change and provides a one-directional flow of information to the body of political decision-makers — the UNFCCC. There are some interesting exceptions to this rule, for example, the fact that the *Summary for Policy-Makers* of each Assessment Report is negotiated jointly by scientists and state representatives and has to be approved by the latter before it is published by the IPCC. But despite these deviations, the IPCC model reflects the conventional way of thinking about the science-policy interface.

While this model has many observed flaws, my main concern here is not with the way the IPCC organizes the provision of scientific information to the diplomatic community. Instead I will assess whether and to what extent diplomats receive and use this information and integrate it into their beliefs and negotiation positions. My focus is not on governance and institutions of science in diplomacy, but on the individual minds of science users. What do they understand, and what do they ignore? What do we need to understand about the mind to understand how it processes scientific information about climate change? What is the role — if any — of emotions in science communication and learning?

Indirectly, answers to these questions speak to the strengths and weaknesses of the institutions at the science-diplomacy interface, and whether and how they should be reformed. If scientific knowledge does not inform political thinking or does so insufficiently, one could point to bad institutions, blame weak substance or lacking communication skills ('Too many numbers, too few charts, too much technical detail, not enough local detail,...'). But one could also argue that limitations of the mind shape the process of absorbing and using scientific information. If that is where the problem lies, reforming the governance of boundary organizations will not fix anything.

Seeking to identify science-related beliefs and positions, I use a cognitive approach. Basic theories about the mind, such as emotional coherence (Thagard, 2006), form the analytic foundation of this research, which conceptualizes the mind as a complex system — a large system of networked concepts and beliefs (mental representations). In these networks, meaning emerges from the presence and absence of links between various nodes (i.e., the network structure), and the parallel activation of these links. In line with recent research in the cognitive sciences, I pay attention to the role of emotions, which is an integral part of human cognition (Vohs *et al.*, 2007; Thagard, 2008; Pessoa, 2008; Roeser, 2011).

Methods to Identify Belief Systems

The main data sources for this research are confidential interviews with climate change negotiators conducted in 2012, supplementary text material, including transcripts of press conferences and speeches of various delegations, and observations of climate negotiation sessions in Durban, South Africa in 2011 and Bonn, Germany in 2012.

The data presented here was gathered for a larger research project on the role of cognition in global climate change politics, conducted in 2012/13. The project's aim was to identify cognitive patterns concerning multilateral cooperation on climate change among participants in the UNFCCC negotiations.

Science-related issues were part of these cognitive patterns, and were subject of two direct interview questions.[1] Other questions exploring participants' general beliefs about cooperation either required or gave the opportunity to use scientific knowledge when formulating an answer.

55 individuals participated in this research; 36 of these were diplomats representing 30 countries and all major negotiation groups.[2] Nineteen individuals represented seven different types of non-governmental organizations (NGOs): youth, faith, development, environment, environment and market, business and technology, fossil fuels.

The selection process was designed to maximize the diversity of viewpoints among study participants. To identify diplomats with diverse perspectives, all countries were categorized into one of six groups based on a combination of their national greenhouse gas (GHG) emissions (high, medium and low) and their climate vulnerability (high, low), using publicly available emissions data and vulnerability indices. Delegation members from countries in each group were contacted until at least two individuals in each group agreed to participate.

Science in the Mind

The interview data reveal a number of surprising patterns concerning the relevance and use of climate science in the belief systems of climate change negotiators. The three sections below address in turn: (a) the general observation that science does not play a significant role in the belief systems of most

[1] (1) How would you summarize the most important and well-established scientific facts about climate change? (2) Which future climate change impacts are you most concerned about? When do you expect serious climate impacts to occur?

[2] Umbrella Group, EU, G77 and China, Least-Developed Countries, Alliance of Small Island States/Small Island Developing States, BASIC (Brazil, South Africa, India, China), Environmental Integrity Group, Bolivarian Alliance for the Americas (ALBA), and Alliance of Rainforest Nations.

study participants — in other words, there is a general disconnect between the science and the politics of climate change; (b) the absence of accurate knowledge about climate tipping points among almost 90% of all study participants; and (c) the common inability to imagine long-term, climatically changed futures.

The science-politics disconnect

Given the nature of climate change as an environmental phenomenon rooted in physical, chemical and biological processes, scientific knowledge should play a major role in the way negotiation participants think about the governance challenge. One would expect that science-based concepts determine or at least influence the general problem definition, possible solution options, the overall goal of climate governance, and maybe even the social and technological challenges to be expected in pursuit of this goal. One would also expect that these various beliefs are dynamic, and change over time based on the latest scientific findings presented in IPCC reports and presentations.

The interview data suggest that with one exception — the goal of limiting global warming to 2°C above pre-industrial levels — these expectations are not being met. Rather than being integrated in the various thought patterns concerning the nature of the governance problem, solution options and governance goal, climate science tends to be mentally separated from these governance and political beliefs. Negotiators' belief systems contain a set of science-based concepts that can be activated or triggered by a specific question, but this cluster does not necessarily interact with the thought processes that create negotiation positions and the ideas that dominate the negotiation process.

Concerning the 2°C target, most negotiators consider this number to be 'imposed by science' and accepted by the diplomatic community. Most scientists and the IPCC disagree with this assessment and emphasize that the selection of a goal is a value judgment that has to be made by political actors (IPCC, 2001). Nevertheless, the majority of study participants believe that this goal is sufficiently ambitious, worth pursuing and technically still achievable. However, in 2012 many already doubted that the temperature goal will be reached, given the political conditions and the speed and scale of change needed to reverse current GHG emission trajectories.

Not only is science-related knowledge poorly integrated in larger political belief systems, it is also not very detailed and it rarely gets updated. All participants share a similar set of scientific beliefs concerning the nature of the problem, which was scientifically established in the first IPCC Assessment Report in 1990. This belief cluster usually contains four major elements: climate change is real, anthropogenic, caused mainly by GHG emissions and land-use change, and has a range of bad consequences. This set of core concepts ('facts') has become a background truth for all political thinking without making reference to numbers, uncertainties, timelines or other specifics. It is taken for granted, does not seem to require articulation anymore, and appears to be a sufficient basis for debating action. This simplified version of climate change — GHG emissions have bad consequences — is a cognitive framework that provides direction, which most individuals are satisfied with despite not knowing how to define success for the climate regime, how to measure it, what would constitute useful milestones, and what relevant time constraints might exist.

The Alliance of Small Island States (AOSIS) is an interesting exception — not only does climate science play a significant role in the beliefs of its delegates, scientific information also informs the bloc's negotiation position, including the demand to limit average global warming to 1.5°C rather than 2°C.

These findings raise challenging questions for rational choice theorists, who assume that negotiation positions are the results of rational costs-benefit calculations and aim to maximize a country's utility. The data suggest that no such calculation takes place. Negotiators do not use scientific data to assess specific costs of climate impacts or benefits of climate policies for their respective countries. Neither do they update such cost estimates and calculations when new scientific data become available, for example, ramping up their willingness to pay for adaptation with increasing cost estimates for damage from sea-level rise. Such calculations might not even be possible with the scientific models and data available today. While some countries' positions might be backed up with numbers from economic models, these numerical results are not part of most negotiators' beliefs.

Ignorance is bliss

One of the most striking features of the interview data is the lack of productive engagement with the idea of tipping points among climate change negotiators.

Over the last few years, climate scientists have pointed out that the global climate and various climate subsystems could exhibit tipping points — 'a critical threshold at which a tiny perturbation can qualitatively alter the state or development of a system" (Lenton *et al.*, 2008: 1786). Examples of systems that could tip include various ice sheets (large and stable ice volume vs. none), the Amazon (rainforest vs. savannah), and the Indian Summer Monsoon (strong vs. weak). While scientific knowledge about tipping points or threshold behaviour at local and regional scales is becoming increasingly sophisticated, scholars are cautious concerning the possibility of global-scale tipping points (Lenton and Williams, 2013), and continue to emphasize a number of major uncertainties concerning the conditions and timing of such events (Lenton, 2012).

Broecker was among the first to express concern about the possibility of climatic tipping points in 1987 (Broecker, 1987). However, the concept tipping point is not new or specific to the climate system. It is often used interchangeably with the terms threshold, regime shift or critical transition. And the concept does not only apply to natural systems like the climate, a forest or a lake, but can also frame one's understanding of social system change. Malcolm Gladwell's book *The Tipping Point* (Gladwell, 2002) popularized the term, although Gladwell's tipping points refer to very different phenomena, such as the spread of a fashion fad, than those that concern Lenton and other scientists, who seek to understand the potentially non-linear dynamics of the climate system. These nonlinear change dynamics are driven by internal system processes such as feedback effects, and can be difficult or impossible to reverse (Lenton *et al.*, 2008; Scheffer, 2009; Scheffer *et al.*, 2009; Lenton, 2011; Dakos *et al.*, 2012).

Tipping as a specific system behaviour could have very serious implications for the wellbeing of human societies, which have evolved in and adapted to the relatively stable climate conditions over the last 10,000 years. The possibility of climate tipping points therefore poses major challenges for the design of climate governance institutions. Gardiner suggests that the growing awareness of the possibility of tipping points should be welcomed because it could undermine the current political inertia and therefore 'help us to act' (Gardiner, 2009: 140). Nuttal also argues that the idea of tipping points has major discursive power and prompts 'discussion characterized by a nervous anticipation of the future' (Nuttall, 2012: 97).

Given the fast-growing scientific interest in climatic tipping points and the significant implications they could have for future human wellbeing, one would expect climate change decision-makers to be interested and possibly concerned about the topic. At a minimum, diplomats should be learning about the concept in order to form an opinion on the relevance of tipping points for climate governance — they need to decide whether anything can and should be done about them.

Based on the interview data, the majority of study participants were neither informed nor concerned about tipping points in 2012. A not insignificant number of participants responded to a question about the relevance of tipping points with the question, 'What do you mean by tipping points?' Only six participants (11 percent) have a good understanding of the concept and complex systems more generally. Three out of these six were diplomats, representing a mere eight percent of all diplomatic participants. Some individuals in this group believe that we might be close to or have passed a tipping point in the Arctic.

In addition to these six well-informed individuals, 15 participants (four diplomats and 11 NGO representatives) mentioned the concept tipping points, but usually only when specifically asked about it. More importantly, the meaning of the term tipping point varied significantly among them, suggesting that most have not yet understood the concept and its implications for climate governance rather than climate science:

- 'We would like to avoid all of the repercussions of reaching a certain threshold, of which many countries think it's 2°C … the 2°C tipping point';
- 'Once you hit those [tipping points], you know things just get out of hand';
- 'All the biological systems work in that way. …You reach the tipping point and then you go straight down. It's never a comfortable sort of landing, it's always like a crash landing. …the whole system falls apart';
- 'But a certain trigger changing in climate would be irreversible and dramatic, trains would be leaving their station, and couldn't just be pulled back';
- 'We don't really know what's going to happen with the climate when the global mean temperatures increase. Some things might happen that we don't, we can't predict and there maybe points where we can't return';

- 'That would be the point where it's becoming so real and bad that perhaps people will start to care';
- 'I guess the tipping point I was always concerned about is missing a date for the peaking of global emissions, because what the scientists are saying is that globally we should peak between 2015 and 2020';
- '...this idea of you can't go back home, you know, you can't go back to where you were before, if you have a tipping point'.

Most definitions converge around the idea that there might be a point in time when things 'get out of hand' and living conditions for people everywhere would get significantly worse without any hope to return to the way things are today.

These observations — the limited engagement with the issue and lack of a shared definition — contradict Nuttall, at least as far the climate negotiations are concerned, who suggests, 'The tipping point thus becomes tremendously powerful in discursive, rhetorical, and metaphorical senses'. (Nuttall, 2012: 97). Neither does the data confirm Bellamy and Hulme (2011), who come to the more differentiated conclusion that different value systems determine the effect of tipping point concerns on individual beliefs about climate change risk and action. Rather than coming to different assessments and conclusions about the importance of tipping points, so far the majority of negotiators do not engage with the topic at all. Those who do tend to lack a clear definition, but are worried about the potential consequences and irreversibility of tipping points.

This situation raises two questions: Why are most climate negotiators not thinking about tipping points? What, if anything, could be done to change this situation? I will focus on the first here, and return to the second in the concluding discussion of this chapter.

Several explanations could be offered for the causes of the tipping point knowledge and concern deficit. One could argue that most negotiators are familiar with the concept but after careful assessment they have decided that it is not worth worrying about tipping points. The quotes presented above suggest that informed lack of concern is not what explains the general lack of attention to the issue. Every negotiator who spoke about tipping points was deeply concerned about their potential impacts on human wellbeing and about the currently insufficient scientific understanding and predictability of tipping points. More importantly, many interviewees inquired what I meant by tipping points, indicating that they were not familiar with the idea.

Using the conventional science-diplomacy model, one could argue that diplomats cannot be expected to be aware of tipping points at this point in time because past IPCC reports did not address this issue. The science on climate tipping points might have emerged only after the publication of the latest Assessment Report in 2007, triggered by a famous paper published in 2008 (Lenton *et al.*, 2008). This argument is hardly convincing. Tipping point science has been around for a much longer time (Scheffer *et al.*, 2001; Niemeyer *et al.*, 2005) and the IPCC Fourth Assessment Report (AR4) mentions tipping points in various places (Pachauri and Reisinger, 2007). Admittedly, the concept does not stand out as an important one and it is not mentioned in the Synthesis Report or the Summary for Policy-Makers. However, if one adopted this explanation, one could ask whether diplomats can be expected to read beyond the IPCC reports, or whether the IPCC should provide special reports or presentations on special topics between the major reports, which are published every five to eight years. One would also expect the latest IPCC Assessment Report (AR5) to place greater emphasis on tipping points and consequently a greater awareness and concern among negotiators after its publication in 2013/14.

I suggest that it is more fruitful to explore the role of cognition and emotion for the limited knowledge and concern about tipping points. The concept of tipping points is new and as six study participants have demonstrated, it can be integrated into individual belief systems through learning. However, there might be cognitive-affective barriers to learning about climate tipping points, even if one receives scientific information about the concept. Below I briefly mention two such barriers. Both require a conception of the mind as a complex system, in which cognition and emotion cannot be separated (Damasio, 1995; Loewenstein *et al.*, 2001; Thagard 2006; Vohs *et al.*, 2007; Pessoa, 2008). One explanation is primarily cognitive; the other primarily emotional.

The first challenge to integrating the notion of tipping points into a person's belief system arises from our standard beliefs about causation and change. Most people's understanding of causation is Humean, based on notions of logical independence, temporal succession and contiguity (Hume, 1738/2010). This is a linear understanding where a cause is temporarily prior to an effect. Tipping points in the sense that Lenton and his colleagues use the term do not fit this causal pattern; they are emergent. Emergence is a complex systems feature that describes how novel phenomena at a macro-scale arise out of multiple and usually fairly simple interactions of system elements at a micro-scale (Goldstein, 1999). Given that linear causation and emergence are two ontologically different models of change, and most people know nothing or very little about emergence, understanding tipping points is very challenging. As Chi argues, misconceptions that involve different ontological categories, in particular causation and emergence, are highly robust to change (Chi and Roscoe, 2002).

The second reason for lacking attention to tipping points might be the human tendency to avoid negative emotions and scary thoughts that create anxiety and a feeling of being out of control. This phenomenon has been described as distancing (Norgaard, 2006; 2011; Spence *et al.*, 2012), and I will return to this issue below.

Both of these arguments suggest that even if individuals engaged with scientific information about tipping points, they would find it difficult to understand the concept and make it part of their active belief system, unless they had prior instruction in complex systems thinking and emergence, and found ways to overcome their emotional resistance to dealing with unsettling information.

Imagine, it is 2080, and...

Climate change is a governance challenge with unique temporal features. Most fundamentally, thinking about climate change requires the ability to imagine the future and to understand how the future is linked to the present. Although the problem bears some resemblances with the long time horizons of managing nuclear weapons and nuclear waste, it requires a more subtle understanding of the linkages between the collective, social and economic behaviour of people around the world over multiple decades, the changes in the climate system and the feedback between the two. Understanding the power of the present generations of human beings to influence the wellbeing of many generations to come not through the use of a specific technology, but mitigated by natural systems, is an unprecedented challenge; yet, it is a fundamental requirement for making 'good' decisions. The long-time scales of climate change processes and the factual reach of today's generation far into the future through climate-related decision-making present the international community with novel types and scales of responsibility and impact. The novelty of this situation also presents an unprecedented cognitive challenge — how can and should we think about it? How are today's choices linked to the future? What kind of future do they make more or less likely? And given that complex systems are unpredictable, what is the value of imagination for decision-making?

Imagining possible futures requires a "translation" of scientific information into "pictures in the mind," images of landscapes and cityscapes, stories of people and their experiences, feelings, smells and sounds of future creatures and places. Our imagination of possible futures, both at the desirable and undesirable ends of the spectrum, has to include complex global environmental change processes, but also the multiple possibilities for social change and transformation — forced or pursued — that are intrinsically linked to changes in nature. The recognition of complexity and consequently unpredictability of the future does not render the future meaningless for present decision-making. Rather than seeking guidance from unrealistic computer models, decision-makers can and should rely on their ability to understand the general direction and kinds of change that are possible and likely, and how their choices interact with and shape this complex system.

But to what extent does the future play a role in the belief-systems of decision-makers today?

The interview data suggest that study participants find it particularly challenging to imagine the distant future. One interview question targeted this issue, asking study participants how they imagined a worst-case scenario of a world where governance efforts had failed and continued GHG emissions had led to the worst climate impacts they could imagine in the year 2080. The reaction to this question showed a surprisingly consistent pattern across all participant groups. With very few exceptions there was a three-stage response. First, people stated that they had never thought about this scenario and found it challenging to come up with a response on the spot. Second, many participants rejected the idea of a worst-case scenario because 'by then we will certainly have solved the problem'. When pushed to accept the hypothetical worst-case world, many made a final attempt to evade the cognitive challenge, arguing that between now and 2080 new technological solutions will emerge, which we cannot even think of today.

The individuals who described a worst-case 2080 scenario fell into two groups. Those who tried to imagine such a world for the first time during the interview tended to develop ideas that were mostly linear extensions of current or expected climate impacts: more floods, more droughts, more hunger and poverty, more extreme weather events, and occasionally more resource conflicts. The small group of individuals, who had considered this issue before, presented an image of an incredibly negative, dark and scary future world with fewer states, more violent conflict, resource scarcity and a lot less happiness than today. Both groups often referred to movies (*'The Day After Tomorrow'*) to help them and I visualize what they imagined.

These cognitive patterns suggest that most participants in climate negotiations have no reason or ability to imagine the distant future, and that their dominant cognitive pattern is the denial of the possibility of governance failure. Given the strong negative emotions associated with the dystopias described by some study participants, one could argue that the avoidance of images of the distant future is a cognitive self-protection mechanism, and might be productive in the sense that it allows people to work on this issue without becoming depressed. On the other hand, the lack of imagination and the rejection of the possibility of failure means that people are never fully cognizant of what is at stake in the negotiations and what they are collectively putting at risk. The ease with which thoughts about potential future damages are suppressed might significantly limit the motivation of negotiators to come to a cooperative agreement today. How can you determine your own willingness to pay if you don't know what your payment will gain or what costs non-payment could imply?

Research in psychology and sociology describes these cognitive patterns as distancing — an active mental process that represents climate change as something that is distant from the individual or the in-group (Norgaard, 2006). Others discuss them as a cognitive dismissal of very hard problems (Wagner, 2012). The literature focuses on risk perceptions regarding present problems (Spence *et al.*, 2011), and has identified four interacting dimensions of distancing: social (i.e., the problem is perceived to concern other groups), geographical (i.e., it is a problem in far-away places), temporal (i.e., it will happen in the future), and distancing based on uncertainty (i.e., it is unclear whether harm will really occur) (Liberman and Trope, 2008; Spence *et al.*, 2012). Since individuals tend to distance themselves from climate change in the present, it is not surprising that distancing effects are even stronger regarding climate change in the long-term future. Events that are not expected to occur in a person's lifetime are naturally less important than those that might have personal relevance. However, in the context of the climate negotiations, this cognitive pattern is counterproductive. Given the major implications of action or inaction of this generation for the effects of climate change on generations to come, negotiation participants bear responsibility for the distant future. They may decide not to value it as highly as the present and near-term future in their decisions, but failure to even consider distant costs is an impediment to good decision-making.

This observation highlights the role of time in climate diplomacy — both the past (memories) and the future (anticipation) can influence decisions and behaviour of political agents in the present. Since mental representations about the past and the future can be associated with different types of emotions, the salience and intensity of these emotions can shape decisions taken in the present in important and surprising ways. Both memories and anticipation can enter a person's thoughts or decision-making process in different ways. In contrast to the certainty and colourful, emotional vividness that our minds can activate with memories of the past, thoughts about the future do not benefit from the cognitive intensity that accompanies experience. Instead, anticipation of the future depends to a large degree on our ability to imagine and visualize things that have not yet come to pass. We have very weak or no affective links to the future.

Social cognitive theory integrates the concept of time in the form of forethought and goal-setting, which are symbolic cognitive activities. Bandura emphasizes that these symbolizing capabilities enable the human species to transcend the present and shape our life circumstances and even 'override environmental influences' (Bandura, 2006: 164). The future cannot be causal for behaviour in the present because it has not happened yet, but its cognitive representation in the present can turn concepts about the future into presently motivating factors (Bandura, 1989: 1179). The capacity to extrapolate future consequences from known facts enables us to take corrective action to avoid future harm, which presumably increases prospects for human survival (Bandura, 1989: 1181).

Not only concepts but also emotions connected with the mental representation of the future matter. Loewenstein and Elster state that people savour anticipated good events ahead of their occurrence, and dread bad ones

(Loewenstein and Elster, 1992). Dreading bad climate impacts — to the extent that they can be anticipated and emotionally experienced — should therefore motivate individuals today to make choices that prevent future harm.

The data gathered for this study suggests that the easiest way for the mind to deal with the challenge of anticipation is to use the experience of the past and extend it into the future, even if there is knowledge that the future will be different than the past. The lack of long-term imagination and the associated weakness or even absence of emotions might prevent humans from using their cognitive survival skills sufficiently. This cognitive challenge affects a person's ability to value the future or to estimate costs expected to occur in the future (Berns and Atran, 2012). The unprecedented nature of climate change and the perceived temporal distance of severe impacts might simply outstrip current cognitive capabilities. There might be evolutionary explanations for the lack of our imaginative abilities — humans simply never had to think that far ahead. Although we have massively expanded our decision-making time horizon from days in the time of hunters and gatherers to multiple years in modern societies, timelines of multiple decades or even a century are simply beyond our grasp at this point in human history.

In a nutshell, these theoretical and empirical insights imply that the past might have a fairly strong influence on decisions today, while certain, more distant parts of the future might not even enter the cost-benefit calculation or the normative considerations of climate negotiators. At present, institutions at the science-diplomacy interface do not acknowledge or address this cognitive dimension that links time — thoughts and emotions related to the past and the future — to decisions and behaviour in the present.

The Future of Science in Diplomacy

This chapter has offered a number of empirical clues about present and future challenges for the increasingly important ties between science and diplomacy. The common lack of accurate knowledge about climate tipping points and the difficulties of imagining long-term climatically changed futures share two commonalities. First, both require a conceptual understanding of non-linear change and emergence — features of complex systems. Second, both issues involve thinking about hypothetical future realities that are qualitatively different than anything humans have experienced in the past. When contrasting reality and possibility, knowledge about the past and present gained through experience tends to overwhelm abstract scientific information about what will be.

This imbalance between available and necessary knowledge places important, yet unnoticed, cognitive constraints on current regime-creation efforts. Limiting the availability of thoughts and concepts related to future climate-related impacts, they limit the motivational power such concepts can and probably should have on present decision-making about climate governance.

While these decision-making challenges might not be surprising from an evolutionary perspective, they are detrimental for climate change governance and potentially the wellbeing of future generations. If one defines diplomatic success in terms of effective environmental governance rather than finding political compromise, diplomatic success in the climate negotiations depends to some extent on a cognitive skill set that does not exist or is at best severely underdeveloped among negotiation participants. This skill set contains scientifically informed imagination of long-term futures, and complex systems thinking.

Current forms of scientific information do not and are not supposed to build these skills. This is a situation where things do not seem to work well, but nobody is at fault. Neither the IPCC nor individual diplomats are responsible for this failure — it is the result of multiple interacting factors, including things as diverse as evolution, cognitive mechanisms, and historically grown approaches to science communication in international relations. So what can be done to address this cognitive skill deficit?

To begin with, a lot could be learned from ongoing research efforts to understand the way the human mind works, changes and learns. Kahan's cultural cognition thesis is a valuable starting point (Kahan *et al.*, 2011; Kahan, 2012). Cultural cognition theory suggests that people rely extensively on cultural meanings when they form risk perceptions or assess risk-related information. Using this perspective, climate science communication in the UNFCCC could benefit from a two-channel approach (Kahan *et al.*, 2012), combining scientific information as currently provided by the IPCC and cultural meanings that allow negotiators from diverse cultural backgrounds to actively engage with new concepts and integrate them into their belief systems.

Taking these insights a step further, current institutions and processes of science communication might require a radically different approach. Instead of focusing on the verbal presentation of scientific findings, it might be worth considering novel forms of increasingly visceral and graphic science communication. Where numbers, data and charts fail to draw attention, visuals, experiences, story and narrative could become unconventional, yet important, tools for science communicators that could engage not only cognitive but also emotional and tactile faculties of decision-makers.

Advancing science in climate change diplomacy does not require more or better data or a different IPCC governance model. The key to successful science communication lies in understanding the minds of science users and helping them imagine the possible worlds their decisions are shaping. Such scientifically grounded imagination is a task that bridges science and politics, and cannot be accomplished by either side alone. Bridge-builders are needed, who can bring scientists and decision-makers together, and link scientific knowledge and beliefs though communication forms and representations that match the strengths of the human mind: learning through story, narrative, experience and emotion.

References

Arimoto, T. and Sato Y. (2012) 'Rebuilding public trust in science for policy-making', *Science*, 337(6099): 1176–1177. DOI: 10.1126/science.1224004.

Bandura, A. (1989) 'Human agency in social cognitive theory', *American Psychologist*, 44(9): 1175–1184. DOI: 10.1037/00030-66X.44.9.1175.

——— (2006) 'Toward a psychology of human agency', *Perspectives on Psychological Science*, 1(2): 164–180. DOI: 10.1111/j.17456-916.2006.00011.x.

Bellamy, R. and Hulme M. (2011) 'Beyond the Tipping Point: Understanding perceptions of abrupt climate change and their implications', *Weather, Climate, and Society*, 3(1): 48–60. DOI: 10.1175/2011WCAS1081.1.

Berns, G.S. and Atran S. (2012) 'The biology of cultural conflict', *Philosophical Transactions of the Royal Society B: Biological Sciences*, 367(1589): 633–639. DOI: 10.1098/rstb.2011.0307.

Broecker, W.S. (1987) 'Unpleasant surprises in the greenhouse?' *Nature*, 328(6126): 123–126. DOI: 10.1038/328123a0.

Chi, M.T.H. and Roscoe R.D. (2002) 'The processes and challenges of conceptual change', in Limón M. and Mason L. (eds.) *Reconsidering Conceptual Change: Issues in Theory and Practice*, Netherlands: Springer, pp. 3–27.

Crutzen, P.J. (2006) 'The "Anthropocene"', in Ehlers E. and Krafft T. (eds.) *Earth System Science in the Anthropocene*, Berlin, Germany: Springer Berlin-Heidelberg, pp. 13–18.

Dakos, V., Carpenter S.R., Brock W.A., Ellison A.M., Guttal V., Ives A.R., Kéfi S., Seekell D.A., van Nes E.H. and Scheffer M. (2012) 'Methods for detecting early warnings of critical transitions in time series illustrated using simulated ecological data', *PLoS ONE*, 7(7): e41010. DOI: 10.1371/journal.pone.0041010.

Damasio, A.R. (1995) *Descartes' Error: Emotion, Reason, and the Human Brain*. 1st edition. New York: Harper Perennial.

Folke, C. (2006) 'Resilience: The emergence of a perspective for social-ecological systems analyses', *Global Environmental Change*, 16(3): 253–267. DOI: 10.1016/j.gloenvcha.2006.04.002.

Funtowicz, S.O. and Ravetz J.R. (1993) 'Science for the post-normal age', *Futures*, 25(7): 739–755. DOI: 10.1016/00163-287(93)90022-L.

Galaz, V. (2013) 'A "Planetary Boundaries" Straw-Man', *Resilience Science*, 8 April, [Online], Available: http://rs.resalliance.org/2013/04/08/a-planetary-boundaries-straw-man/ [21 May 2014].

Gamson, W.A. (1999) 'Beyond the science-versus-advocacy distinction', *Contemporary Sociology*, 28(1): 23–26. http://www.jstor.org/stable/2653844.

Gardiner, S.M. (2009) 'Saved by Disaster? Abrupt climate change, political inertia, and the possibility of an intergenerational arms race', *Journal of Social Philosophy*, 40(2): 140–162. DOI: 10.1111/j.14679-833.2009.01444.x.

Gladwell, M. (2002) *The Tipping Point*. New York: Back Bay.

Goldstein, J. (1999) 'Emergence as a Construct: History and issues', *Emergence*, 1(1): 49–72. DOI: 10.1207/s15327000em0101_4.

Ho-Lem, C., Zerriffi H. and Kandlikar M. (2011) 'Who Participates in the Intergovernmental Panel on Climate Change and Why: A quantitative assessment of the national representation of authors in the intergovernmental panel on climate change', *Global Environmental Change*, 21(4): 1308–1317. DOI: 10.1016/j.gloenvcha.2011.05.007.

Hoppe, R., Wesselink A. and Cairns R. (2013) 'Lost in the Problem: The role of boundary organisations in the governance of climate change: role of boundary organizations in the governance of climate change', *Wiley Interdisciplinary Reviews: Climate Change*, 4(4): 283–300. DOI: 10.1002/wcc.225.

Hulme, M. (2009) *Why We Disagree About Climate Change: Understanding Controversy, Inaction and Opportunity*. 1st edition. Cambridge: Cambridge University Press.

Hulme, M. and Mahony M. (2010) 'Climate Change: What Do We Know About the IPCC?' *Progress in Physical Geography*, 34(5): 705–718. DOI: 10.1177/0309133310373719.

Hume, D. (1738/2010) *A Treatise of Human Nature*. Available: http://www.gutenberg.org/files/4705/4705-h/4705-h.htm.

International Panel on Climate Change (IPCC) (2001) 'Climate Change 2001: Impacts, adaptation and vulnerability', *IPCC Third Assessment Report*, Available: http://www.grida.no/publications/other/ipcc_tar/.

Jasanoff, S. (2004) *States of Knowledge: The Co-production of Science and the Social Order*, New York: Routledge.

Kahan, D. (2012) 'Cultural cognition as a conception of the cultural theory of risk', in Roeser S., Hillerbrand R., Sandin P. and Peterson M. (eds.) *Handbook of Risk Theory*, Netherlands: Springer, pp. 725–759.

Kahan, D., Jenkins-Smith H.C., Tarantola T., Silva C.L. and Braman D. (2012) 'Geoengineering and the Science Communication Environment: A Cross-Cultural Experiment', *SSRN Scholarly Paper ID 1981907*, Rochester, NY: Social Science Research Network. DOI: 10.2139/ssrn.1981907.

Kahan, D., Jenkins-Smith H.C. and Braman D. (2011) 'Cultural cognition of scientific consensus', *Journal of Risk Research*, 14(2): 147–174. DOI: 10.1080/13669877.2010.511246.

Lackey, R.T. (2007) 'Science, scientists, and policy advocacy', *Conservation Biology*, 21(1): 12–17. DOI:10.1111/j.15231-739.2006.00639.x.

Lenton, T.M. (2012) 'Arctic climate tipping points', *AMBIO: A Journal of the Human Environment*, 41(1): 10–22. DOI: 10.1007/s132800-110-221-x.

Lenton, T.M. (2011) 'Early warning of climate tipping points', *Nature Climate Change*, 1 (4): 201–209. DOI: 10.1038/nclimate1143.

Lenton, T.M., Held H., Kriegler E., Hall J.W., Lucht W., Rahmstorf S. and Schellnhuber H.J. (2008) 'Tipping elements in the earth's climate system', *Proceedings of the National Academy of Sciences*, 105(6): 1786–1793. DOI: 10.1073/pnas.0705414105.

Lenton, T.M. and Williams H.T.P. (2013) 'On the origin of planetary-scale tipping points', *Trends in Ecology and Evolution*, 28(7): 380–382. DOI: 10.1016/j.tree.2013.06.001.

Levin, K., Cashore B., Bernstein S. and Auld G. (2012) 'Overcoming the tragedy of super wicked problems: constraining our future selves to ameliorate global climate change', *Policy Sciences*, 45(2): 123–152. DOI: 10.1007/s110770-012-9151-0.

Liberman, N. and Trope Y. (2008) 'The Psychology of Transcending the Here and Now', *Science*, 322(5905): 1201–1205. DOI: 10.1126/science.1161958.

Loewenstein, G.F. and Elster J. (eds.) (1992) *Choice over Time*, 1st edition, New York: Russell Sage Foundation.

Loewenstein, G.F., Weber E.U., Hsee C.K. and Welch N. (2001) 'Risk as feelings', *Psychological Bulletin*, 127(2): 267–286. DOI: 10.1037/00332-909.127.2.267.

Miller, C.A. (2007) 'Democratization, international knowledge institutions, and global Governance', *Governance,* 20(2): 325–357. DOI: 10.1111/j.14680-491. 2007.00359.x.

Niemeyer, S., Petts J. and Hobson K. (2005) 'Rapid Climate Change and Society: Assessing responses and thresholds', *Risk Analysis,* 25(6): 1443–1456. DOI: 10.1111/j.15396-924.2005.00691.x.

Norgaard, K.M. (2006) 'People Want to Protect themselves a Little Bit: Emotions, denial, and social movement nonparticipation', *Sociological Inquiry,* 76: 372–396. DOI: 10.1111/j.14756-82X.2006.00160.x.

——— (2009) 'Cognitive and behavioral challenges in responding to climate change', *World Bank Policy Research Working Paper Series,* SSRN eLibrary, http://papers.ssrn.com/sol3/papers.cfm?abstract_id=1407958.

——— (2011) *Living in Denial: Climate Change, Emotions, and Everyday Life,* 1st edition, Cambridge, MA: MIT Press.

Nuttall, M. (2012) 'Tipping Points and the Human World: Living with Change and Thinking About the Future', *AMBIO: A Journal of the Human Environment,* 41(1): 96–105. DOI: 10.1007/s13280-011-0228-3.

Pachauri, R.K. and Reisinger A. (2007) 'Climate Change 2007: Synthesis Report — Contribution of Working Groups I, II and III to the Fourth Assessment Report of the Intergovernmental Panel on Climate Change', *IPCC Assessment Reports.* Geneva, Switzerland: Intergovernmental Panel on Climate Change. Available: http://www.ipcc.ch/publications_and_data/publications_ipcc_fourth_assessment_report_synthesis_report.htm.

Parker, J.N. and Hackett E.J. (2012) 'Hot spots and hot moments in scientific collaborations and social movements', *American Sociological Review,* 77(1): 21–44. DOI: 10.1177/0003122411433763.

Parker, J.N., Lortie C. and Allesina S. (2010) 'Characterizing a Scientific Elite: The social characteristics of the most highly cited scientists in environmental science and ecology', *Scientometrics,* 85(1): 129–143. DOI: 10.1007/s11192-010-0234-4.

Pessoa, L. (2008) 'On the relationship between emotion and cognition', *Nature Reviews Neuroscience,* 9(2): 148–158. DOI: 10.1038/nrn2317.

Pielke Jr., R. (2013) 'Planetary boundaries as power grab', *Roger Pielke Jr.'s Blog,* 4 April, [Blog], Available: http://rogerpielkejr.blogspot.com/2013/04/planetary-boundries-as-power-grab.html [21 May 2014].

Rittel, H.W.J. and Webber M.M. (1973) 'Dilemmas in a general theory of planning', *Policy Sciences,* 4(2): 155–169. DOI: 10.1007/BF01405730.

Roeser, S. (2011) *Moral Emotions and Intuitions,* 1st edition, Basingstoke: Palgrave Macmillan.

Scheffer, M. (2009) *Critical Transitions in Nature and Society,* Princeton: Princeton University Press.

Scheffer, M., Bascompte J., Brock W.A., Brovkin V., Carpenter S.R., Dakos V., Held H., van Nes E.H, Rietkerk M. and Sugihara G. (2009) 'Early-warning signals for critical transitions', *Nature,* 461(7260): 53–59. DOI: 10.1038/nature08227.

Scheffer, M., Carpenter S., Foley J.A., Folke C. and Walker B. (2001) 'Catastrophic Shifts in Ecosystems', *Nature,* 413(6856): 591–596. DOI: 10.1038/35098000.

Spence, A., Poortinga W., Butler C. and Pidgeon N.F. (2011) 'Perceptions of climate change and willingness to save energy related to flood experience', *Nature Climate Change,* 1: 46–49. DOI: 10.1038/nclimate1059.

Spence, A., Poortinga W. and Pidgeon N. (2012) 'The Psychological Distance of Climate Change', *Risk Analysis,* 32(6): 957–972. DOI: 10.1111/j.1539-6924.2011.01695.x.

Steffen, W., Crutzen P.J. and McNeill, J.R. (2007) 'The Anthropocene: Are humans now overwhelming the great forces of Nature', *AMBIO: A Journal of the Human Environment,* 36(8): 614–621. DOI: 10.1579/0044-7447(2007)36[614:TAAHNO]2.0.CO;2.

Thagard, P. (2006) *Hot Thought: Mechanisms and Applications of Emotional Cognition. Cambridge,* MA: MIT Press.

——— (2008) How Cognition Meets Emotion: Beliefs, Desires and Feelings and Neural Activity', In Brun G., Doguolu U. and Kuenzle D. (eds.) *Epistemology and Emotions,* 1st edition, Aldershot: Ashgate Publishing, pp. 167–184.

Vasileiadou, E., Heimeriks G. and Petersen, A.C. (2011) 'Exploring the Impact of the IPCC Assessment Reports on Science', *Environmental Science and Policy,* 14(8): 1052–1061. DOI: 10.1016/j.envsci.2011.07.002.

Verweij, M., Douglas M., Ellis R., Engel C., Hendriks F., Lohmann S., Ney S., Rayner S. and Thompson M. (2006) 'Clumsy Solutions for a Complex World: The Case of Climate Change', *Public Administration,* 84(4): 817–843. DOI: 10.1111/j.1540-8159.2005.09566.x-i1.

Vohs, K. D., Baumeister R.F. and Loewenstein G.F. (2007) *Do Emotions Help Or Hurt Decision Making?: A Hedgefoxian Perspective.* New York: Russell Sage Foundation.

Wagner, G. (2012) 'Climate Policy: Hard Problem, Soft Thinking', *Climatic Change,* 110(3–4): 507–521. DOI: 10.1007/s10584-011-0067-z.

Chapter 7

The Emperor's New Clothes: A Failure of Diplomacy in the Oil and Mining Sectors

Sefton Darby

A Career of Failure

Fashion dictates disclaimers — so herewith a few. This is not a work based on an extensive reading of the academic literature or crunching of the numbers. It is perhaps better to think of it as a reflection on some years spent as a participant in a somewhat obscure locale of the diplomatic landscape. My perspective changed over the years from that of a government official to an employee of an international financial institution, before I became a consultant to governments, international organisations, and non-governmental organisations (NGOs). That view was also shaped by many levels of interactions — from multi-stakeholder[1] international negotiations to create and build a new international initiative, to negotiations with developed and developing country governments at the ministerial and official level, to working with NGOs and governments at the sub-national (states, districts) level. This is an attempt to give some order and structure to personal experience. In the process, I hope to illustrate a failure of diplomacy that has come about because of the inability of those involved to understand the complexities of the industries concerned and how they relate to one-another and to consumers.

Over the past decade a significant effort has been put into natural resource governance issues in developing countries, with a particular focus on

[1]The term 'multi-stakeholder' is a clunky one. In this chapter it is used as shorthand for meaning 'governments, civil society, and the private sector'.

the oil and mining sectors. That effort has been international, national and local in its focus. It has gone into battle for a noble principle: that if managed well, the successful development of extractive (i.e., oil and mineral) resources can be a game changer when it comes to alleviating poverty in developing countries. If poorly managed, however, a combination of factors can leave resource-rich developing countries susceptible to poverty, authoritarianism and conflict — a phenomenon known as the 'resource curse'.

The premise of this brief tour through the issue is twofold, both of which relate to an underlying failure to fully understand the total 'ecosystem' of international hydrocarbon and minerals extraction and consumption. Firstly, virtually all of the effort that has been expended on resource curse issues by diplomats, aid and development officials, and the NGOs lobbying them, have largely ignored some of the most important players that occupy important niches in the extractives 'food chain'. State-owned and unlisted privately owned corporations have remained well camouflaged throughout the decade of efforts on the issue, watching on while concerned nations and groups have focused their attention on publicly-listed multinational corporations. Secondly, the role of developed country consumers has been completely overlooked by those same diplomats, development officials and NGOs. On this issue the global community faithfully saunters after cliché and is definitely very local, but in the diplomatic milieu the resource curse and solutions to it are things that are almost entirely conceived of as issues of remote importance. These two failures collectively ensure that while an extraordinary amount of diplomacy has been expended on the issue, very little science has.

The Resource Curse

To understand the debate we must understand the issue and the global corner that it has crawled out of. Much has been written both for and against the concept of the resource curse and it is a debate that would happily fill the length of this book and then grumpily go looking for a library to occupy. The resource curse theory claims that countries with a high level of dependence on revenue from the extractive industries tend to be more susceptible to poverty, poor governance and conflict. Figure 1 illustrates this by comparing the positive and negative ends of several indices — the United Nations Development Programme's Human Development Index

Highly developed countries
(resource rich countries highlighted)

HDI	Country	HDI	Country
1	Norway	14	Austria
2	Australia	15	Spain
3	Iceland	16	Denmark
4	Canada	17	Belgium
5	Ireland	18	Italy
6	Netherlands	19	Liechtenstein
7	Sweden	20	New Zealand
8	France	21	United Kingdom
9	Switzerland	22	Germany
10	Japan	23	Singapore
11	Luxembourg	24	Hong Kong
12	Finland	25	Greece
13	United States		

Source: United Nations Development Program, Human Development Report 2009.

Least developed countries
(resource rich countries highlighted)

HDI	Country	HDI	Country
158	Nigeria	171	Ethiopia
159	Togo	172	Mozambique
160	Malawi	173	Guinea-Bissau
161	Benin	174	Burundi
162	Timor Leste	175	Chad
163	Cote d'Ivoire	176	D.R. Congo
164	Zambia	177	Burkina Faso
165	Eritrea	178	Mali
166	Senegal	179	C.A.R.
167	Rwanda	180	Sierra Leone
168	Gambia	181	Afghanistan
169	Liberia	182	Niger
170	Guinea		

Source: United Nations Development Program, Human Development Report 2009.

Least corrupt countries
(resource rich countries highlighted)

CPI	Country	CPI	Country
1	New Zealand	14=	Germany
2	Denmark	14=	Ireland
3=	Singapore	16	Austria
3=	Sweden	17=	Japan
5	Switzerland	17=	United Kingdom
6=	Finland	19	United States
6=	Netherlands	20	Barbados
8=	Australia	21	Belgium
8=	Canada	22=	Qatar
8=	Iceland	22=	Saint Lucia
11	Norway	24	France
12=	Hong Kong	25	Chile
12=	Luxembourg		

Source: Transparency International Corruption Perceptions Index 2009

Most corrupt countries
(resource rich countries highlighted)

CPI	Country	CPI	Country
154=	Paraguay	168=	Eq. Guinea
154=	Yemen	168=	Guinea
158=	Cambodia	168=	Haiti
158=	C.A.R.	168=	Iran
158=	Laos	168=	Turkmenistan
158=	Tajikistan	174	Uzbekistan
162=	Angola	175	Chad
162=	Congo-B	176=	Iraq
162=	D.R. Congo	176=	Sudan
162=	Guinea-Bissau	178	Myanmar
162=	Kyrgyzstan	179	Afghanistan
162=	Venezuela	180	Somalia
168=	Burundi		

Source: Transparency International Corruption Perceptions Index 2009

"Free" countries (resource rich highlighted)

Andorra	Denmark	Luxembourg	St Lucia
Australia	Dominica	Malta	San Marino
Austria	Estonia	Marshall Islands	Slovakia
Bahamas	Finland	Micronesia	Slovenia
Barbados	France	Nauru	Spain
Belgium	Germany	Netherlands	Sweden
Canada	Hungary	New Zealand	Switzerland
Cape Verde	Iceland	Norway	Tuvalu
Chile	Ireland	Palau	United Kingdom
Costa Rica	Kiribati	Poland	United States
Cyprus	Liechtenstein	Portugal	Uruguay
Czech Republic	Lithuania	St Kitts & Nevis	*All score "1" on Freedom House 'Freedom in the World' 2010 survey.*

"Not free" countries (resource rich highlighted)

5.5	Oman	Swaziland	Zimbabwe
Afghanistan	Qatar	Tunisia	**7.0**
Algeria	Russia	Vietnam	Burma
Angola	Rwanda	**6.5**	Eq. Guinea
Azerbaijan	Tajikistan	Belarus	Libya
Brunei	U.A.E.	Chad	North Korea
Cambodia	**6.0**	China	Somalia
Congo-B.	Cameroon	Cuba	Sudan
Cote d'Ivoire	D.R. Congo	Eritrea	Turkmenistan
Egypt	Guinea	Laos	Uzbekistan
Kazakhstan	Iran	Saudi Arabia	*All score between 5.5 – 7.0 on Freedom House 'Freedom in the World' 2010 survey.*
Mauritania	Iraq	Syria	

Figure 1. The resource curse in numbers.

Sources: 2009 indices from United Nations Development Programme, Human Development Index, available at http://hdr.undp.org/en/statistics/hdi/; Transparency International, Corruption Perceptions Index, Available at http://www.transparency.org/research/cpi; and Freedom House, Freedom in the World, Available at http://www.freedomhouse.org/report-types/freedom-world.

(HDI), Transparency International's Corruption Perceptions Index (CPI), and Freedom House's measures of political rights and civil liberties. What these tables show is that countries which are 'resource rich'[2] tend to be less 'developed', more corrupt, and less free.

It is important to qualify the term 'resource rich' — extractive resources in any quantity do not automatically predispose a country to a life of misery. It is only the point at which those resources become macro-critical that one begins to see economic and political distortions. Perhaps a better measure still is the resource endowment per capita. Keith Myers (2005) makes the interesting argument that countries with a high amount of resources per capita (e.g., many of the Gulf states) derive large economic benefits from such an endowment, while the worst possible situation is one in which there is a significant resource endowment spread across a country with a large population with few other industries — i.e., a resource large enough to be worth stealing, to fight over, or to use to build palaces, but not enough to lift people out of poverty.[3] Beyond the actual quantities of resources, the ability to derive long-term and sustainable development from a finite resource depends on a wide variety of factors, from investment decisions to political and economic culture.[4] Enough countries have successfully passed through a resource-driven phase of growth, or are currently ably navigating the challenges that large quantities of resources bring, to suggest that the resource curse is by no means inevitable given adequate oversight and governance.

---

[2]The International Monetary Fund (IMF, 2007: 4) defines a country as being 'resource rich' if it meets 'either of the following criteria: (1) an average share of hydrocarbon and/or mineral fiscal revenues in total fiscal revenue of at least 25 percent...; or (2) an average share of hydrocarbon and/or mineral export proceeds in total export proceeds of at least 25 percent...'.

[3]Nigeria is the obvious example of a country that is often described as being 'resource rich' but which on a per capita basis is actually resource poor. In 2010 it produced an average of 2.4 million barrels of oil per day — a per capita resource endowment of approximately 5.5 barrels of oil per year for each of its 158 million citizens. This compares with Saudi Arabia's 10.5 million barrels per day — which equates to 142 barrels per year for each of its 27 million citizens. A further comparison yet would be with New Zealand's oil production which in the same year averaged 52,882 barrels per day, or 4.4 barrels per person — only slightly less than that of Nigeria.

[4]It is worth noting that this book is a distant beneficiary of such investment decisions and political-economic culture — the founding of the University of Otago in 1869 was partially made possible by the wise investment of some of the wealth generated by the Otago gold rush of the 1860s.

A Tale of Two Pipelines

The resource curse is not a new phenomenon. Indeed, from the 1970s to the 1990s two of the best examples could be found in the South Pacific in Nauru and Bougainville.[5] But during the 2000s, against a backdrop of rising commodity prices, as well as an increased focus on corruption as an issue in aid and development programmes, international development organisations (both bilateral and multilateral) became increasingly concerned about both the opportunity posed by natural resource revenues in developing countries, as well as the potential negative consequences. The origins of those concerns and the driving forces for international intervention in the area of resource governance were mixed but can be roughly attributed to the following four factors (the mix of which varied from state to state, organisation to organisation):

- Energy security: Some states were driven by concerns that unstable authoritarian governments were a threat to the stability of energy flows, that internal disruption in only one or two energy exporters could threaten supplies or massively increase the cost of said supplies.[6]
- Conflict: Others were motivated by the massive humanitarian cost (and potential financial cost of any intervention) of resource-fuelled conflicts. At the time there was plenty of evidence of natural resources being used to sustain conflicts in Angola, the Democratic Republic of Congo (DRC), Liberia and Sierra Leone. Resources remain a crucial factor in the on-going conflicts in the DRC today.
- Impact on aid programmes: Some donor states and organisations saw the potential for massive resource revenues to make aid irrelevant in some countries. This was seen as both a positive and a negative driver.

---

[5]The spectacular mismanagement of the Nauru Phosphate Royalties Trust is a case study of massive resource wealth being squandered extremely quickly. The Bougainville conflict in Papua New Guinea during the 1980s and 1990s was driven by a complex mixture of factors, many of which had their source in the massive wealth that was generated by the Panguna copper mine, and the failure to translate that wealth into local benefits.

[6]This was a particular concern for the UK during the mid-2000s as it shifted from being an exporter of energy to being an importer of energy. Energy security issues were certainly a significant driver behind the Foreign Office's active support for the Extractive Industries Transparency Initiative (EITI), even though the programme was being driven by another part of Whitehall (i.e., Department for International Development (DFID)).

Some organisations saw the potential for resource-rich developing countries to use those revenues to address poverty and to move beyond aid dependency, but others were concerned that such revenues could also undermine any attempts at promoting good governance in developing countries.[7]

- Complicity in corruption: Some bilateral donors were also concerned about the potential for resource-revenue-fuelled corruption to be caused by, or to infect, oil and mining companies headquartered in those donor countries.

Those same international development organisations had good cause to be concerned, as many were involved in financing both small- and large-scale oil and mining projects in developing countries. The early tangible expression of all of these factors came with the advent of two major oil pipeline projects: the Chad-Cameroon pipeline used to export oil from Chad, and the Baku-Tblisi-Cheyhan (BTC) pipeline used to export Azerbaijani oil to Turkey's southern coast, thus avoiding the political and geographical bottlenecks of Russia and the Bosphorus and Dardanneles straights. The International Finance Corporation (the private sector lending arm of the World Bank) was a participant in the financing for both projects, and the European Bank for Reconstruction and Development (EBRD) helped to finance the BTC project. The boards of both of these development finance organisations consist of representatives from the finance and/or international development ministries of the major donor countries.

The primary locations of both of these projects — the countries of Chad and Azerbaijan — had five notable things in common in addition to their oil fields. First, in both cases the full value of their oil fields could only be realised through a major external investment in transportation (i.e., pipelines) in order to get the oil to market. Both countries were extremely poor,[8] both were (and continue to

---

[7]In this regards the aid and development community can be somewhat two-faced, with officials both praising the reduction in dependence on aid while at the same time bemoaning their loss of influence over policy in developing countries.

[8]In 2000 the GDP per capita (current prices, US$) for Azerbaijan and Chad were $648 and $146 respectively. By 2009 (i.e., once both pipelines are up and running) that had risen dramatically to $4,798 and $712 (IMF, 2012).

be) ruled by authoritarian governments,[9] both were/are very corrupt,[10] and both have a history of internal and external conflict.[11] As a result, the conditions surrounding such development lending came under immense scrutiny. During 2003 and 2004 the British Secretary of State for International Development[12] received considerable correspondence relating to both pipelines from various NGOs, and a number of questions on the projects were asked in Parliament.[13]

This scrutiny led to a variety of responses, many of which focused on trying to improve the transparency of, and accountability for, the revenues that would flow from such projects. At a high level, the issue of transparency in the extractive industries has been a subject for discussion at every G8

---

[9]Chad has been ruled by General Idris Deby since 1990. Azerbaijan became independent of the Soviet Union in 1991, and with the exception of the first two years, has been ruled by members of the Aliev family — initially former KGB General and Soviet Politburo member Heydar Aliev, and since 2003 — his son Ilham. Freedom House's annual *Freedom in the World* report assesses levels of political rights and civil liberties globally and provides a score from 1–7 in each area to each country, with 1 being 'most free' and 7 being 'least free'. In the 2012 report Chad scores 7 on political rights, and 6 on civil liberties. Azerbaijan scores 6 and 5 on the two measures (Freedom House, 2012).

[10]From Transparency International, *Corruption Perceptions Index* (CPI). The CPI annually ranks the apparent level of corruption countries around the world. The 2011 CPI measured and ranked 182 countries from 1st (least corrupt- New Zealand) to 182nd (North Korea and Somalia). Chad currently comes in at 168th and Azerbaijan at 143rd (CPI, 2011).

[11]Chad has a long history of civil war and internal conflict and also fought an on-again-off-again war against Libya during the 1980s. Azerbaijan suffered a period of internal instability shortly after independence and fought a war with Armenia over the Nagorno-Karabagh region in the early-1990s.

[12]It was a tumultuous year for the DFID as it went through three Secretaries of State — Claire Short (who resigned over the Iraq war), Valerie Amos and then Hilary Benn. The Extractive Industries Transparency Initiative (EITI) was championed by Claire Short during her time as Secretary of State for International Development, and she eventually returned to the Initiative when she was selected to be the Chair of the EITI Board in 2011.

[13]The weight of correspondence received on this issue did, for a short while, divert some civil servant resources away from actually developing EITI as it emerged. There is an interesting phenomenon here that those engaged in campaigning for a change of government policy should be aware of. Too little scrutiny and accountability runs the risk of poor policy, but too much of it can create a self-fulfilling prophesy — a government department that fails to address an important issue because it is so preoccupied with dealing with correspondence and lobbying efforts on it.

Summit since 2003.[14] The World Bank Group (including the IFC) was subject to an extensive review of its extractive industries-related operations and lending.[15] A coalition of international civil society groups launched the Publish What You Pay (PWYP) campaign in 2002[16] lobbying for greater transparency and accountability, both in 'home countries' (i.e., countries where major oil and mining companies are headquartered) and in developing countries. The concrete manifestation of such pressure was the establishment of the Extractive Industries Transparency Initiative (EITI).[17] It is from this particular nook of the international system that some of the conclusions in this chapter are drawn.[18]

---

[14]References to the various G8 declarations can be found at the G8 Information Centre — http://www.g8.utoronto.ca/summit/index.htm. Multiple G8 declarations should be taken as an indication of an issue that is important enough to be of concern/interest to several member states, but it should not in any way be taken to mean a commitment to collective action. Indeed, at this level of negotiation the effort that goes into detailed communique drafting is wholly disproportionate to the benefits that such efforts actually deliver. At worst these efforts are an employment scheme for diplomats and development officials; at best they slightly assist in internal resource allocation — G8 declarations on EITI helped a small number of civil servants in a small number of countries make a case for continued funding of EITI efforts, but it certainly did not lead to any spontaneous outbreaks of commitments to greater transparency in Moscow or Rome.

[15]The Extractive Industries Review — for the World Bank's Management Response to the EIR see http://www.ifc.org/eir.

[16]Publish What You Pay was launched by Global Witness, CAFOD, the Open Society Institute, Oxfam GB, Save the Children UK, and Transparency International UK. See http://www.publishwhatyoupay.org/.

[17]The Extractive Industries Transparency Initiative (EITI) was launched by the UK Government in 2002 and after a somewhat faltering start eventually became a major international initiative on resource revenue transparency. Initially hosted by the DFID, the EITI is now a fully independent international organisation, headquartered in Oslo and overseen by a board consisting of government, civil society, and company/investor representatives. See http://www.eiti.org.

[18]It should be noted that the EITI is but one of several international voluntary initiatives or standards focused on how multinational companies interact with and operate in developing countries. Examples of other initiatives focused on the extractive industries include The Voluntary Principles on Security and Human Rights (see http://www.voluntaryprinciples.org) and Kimberly Process on conflict diamonds (see http://www.kimberleyprocess.com). The distinguishing feature of such initiatives tends to be whether they are government-driven, industry-driven, or civil-society driven. EITI is more of a genuinely multi-stakeholder initiative/standard than most. It also draws strength from the fact that its international negotiations and standard-setting is matched by very significant country-level implementation.

The rationale at the time was that deals such the Chad-Cameroon and BTC pipelines would happen with or without the participation of development banks such as the IFC and the EBRD. It was expected that in countries such as Chad and Azerbaijan, the revenues generated by the projects would be fundamentally transformative — if managed well they could drive long-term poverty-reduction; if managed poorly they could entrench authoritarian rule, corruption, and sustain or spur new conflict. Involvement in the financing of such deals gave the development organisations a modicum of influence over the situation through which they could promote initiatives such as the EITI.

Bickering at the Margins

In the aid and development world, every proposed solution to an issue has at least two economists and an NGO arguing against it, and the resource revenue transparency movement is no exception to this. The most consistent criticism has been that it focuses on only one link in the extractive industries 'value chain'. These criticisms focus on the need for a broader range of interventions before and after the generation of revenues. The 'before' interventions focus primarily on how permits and contracts for resource extraction are awarded, while the 'after' interventions focus on how the revenues should be saved or distributed/spent. Prominent amongst these comprehensive approaches are the Natural Resource Charter[19] and the World Bank's on-going technical assistance to developing country governments on all of these issues.[20]

These broader approaches are of course right. There is no magic bullet when it comes to avoiding or ameliorating the resource curse. But those who object to the resource revenue transparency movement for this particular

[19]The Charter been driven by economist Paul Collier, see http://www.naturalresourcecharter.org/.

[20]The adoption of the *EITI Standard* by the EITI Board and Conference in May 2013 also signifies an important step by the initiative into other parts of the extractives value chain. To the credit of those involved in EITI it is an initiative that has always tried to evolve its standard at a rate that leads to more and more disclosure, while at the same time preserving the delicate coalition of governments, companies and NGOs that support it — hence why it has taken a decade since EITI was first launched to get to a point where it is able to talk about contract transparency.

reason tend to rely on the 'intellectually purist' argument that one good programme is no good because it is not accompanied by every other necessary programme. It is an approach that runs head on into the political reality of the oil and mining sectors in many developing countries, in which how permits are allocated and how the dividends are distributed are often the most sensitive and guarded features of an economy. The resource revenue transparency movement has gained a lot of traction precisely because it is, on the face of it, relatively inoffensive and unthreatening to corrupt regimes or companies. In many countries implementing the EITI, for example, the programme began with a focus on revenue transparency and, after several years, started asking more probing questions around resource governance. In this regard EITI is a Trojan Horse initiative, able to get through the walls precisely because it looks harmless.

Those obsessed with the savings/investment/distribution debate are also right. In some countries, resource-based sovereign wealth funds are used for a variety of purposes — budget smoothing (saving during times of high commodity prices in order to provide a budget top-up during times of low commodity prices); economic diversification (investment in non-extractive sectors of the economy); and inter-generational wealth distribution (ensuring that future generations will benefit from resource wealth, not just the generations doing the extracting). Collectively these funds have become a significant component of the global economy.[21]

That said, the existence of these funds guards against only some of the economic risks associated with resource dependence. They have not contributed to any great wave of democratisation, nor have they prevented one of most consistently featured economic failures in such states — the wasting of resources through economic and literal monument building. The pouring of billions of dollars into pet commercial projects and half-empty mirror glass towers and grandiose prestige buildings is one of the most consistent features

[21] By way of comparison the funds in Table 1 collectively add up to somewhat more than the annual GDP of France.

Table 1. Resource-based sovereign wealth funds.[22]

Country	Value	LMI
Norway	$737bn	10
Saudi Arabia	$676bn	4
UAE–Abu Dhabi	$627bn	5
Kuwait	$386bn	6
Russia	$176bn	5
Qatar	$115bn	5
Algeria	$77bn	1
UAE–Dubai	$70bn	4
Kazakhstan	$69bn	8
UAE–Abu Dhabi	$65bn	9
Libya	$65bn	1
Iran	$54bn	5

of corrupt oil states, and the illustrations here are but a sample of the petro-architecture one finds in such countries (see Figure 2).[23]

[22]Collectively these funds account for 12 out of the 20 largest sovereign wealth funds in the world, with the notable non-oil ones being a number of Chinese and Singaporean investment funds. The LMI score — the Linaburg Maduell Index — is out of 10 (1 low, 10 high) — and is based on a basket of measures around issues such as publication of information about fund holdings and governance. Abu Dhabi has multiple wealth funds with different objectives and governance arrangements, hence its multiple appearances in the above table. All data is from the Sovereign Wealth Fund Institute rankings, available at http://www.swfinstitute.org/fund-rankings/.

[23]If such buildings are not generated by corruption, then they are at least facilitated by an absence of political accountability that allows elites to construct such surreal edifices with no fear that the wider population might suggest that the funds could be put to better use. The opposite of this phenomenon is also true — the public sector architecture in democracies is often cost-effective and dull precisely because of the fear of public outrage.

Figure 2. Petro-architecture in Kazakhstan's new capital city of Astana

Top left: The Palace of Peace and Reconciliation. Top right: The Bayterek tower. Bottom: President Nursultan Nazabaev's handprint at the viewing platform of Bayterek. All photos copyright Sefton Darby, 2012.

The Multinational Monsters Under the Bed

Within the revenue transparency arena there has long been a debate on whether 'mandatory' or 'voluntary' approaches are best. The terminology is clumsy, but in this context 'mandatory' refers to approaches that would regulate companies in their home countries and require them to report on payments to governments in every country in which they have operations. It was this approach that was the founding principle of the Publish What You Pay movement and which belatedly should have been realised in 2010 with the passing by the US Congress of Section 1504 of the Dodd-Frank Act that requires such mandatory reporting of all companies listed with the Securities and Exchange Commission. The European Commission and Parliament are currently considering similar proposals.[24]

So-called 'voluntary' approaches, on the other hand, leave it up to individual companies as to how much or how little they disclose. The EITI has adopted a middle approach that is best described as 'local mandatory'; while not requiring EITI supporting companies to disclose payments to governments in all the countries in which they operate, it does require all companies operating in an EITI-implementing country to disclose their payments to the government. This approach has frustrated some NGOs, who see little point in a standard that does not mandate disclosure by multinational corporations, but it has achieved something that virtually no other initiative can claim — reporting of payments to governments by state owned companies, privately owned companies, and multinational companies headquartered in countries not normally predisposed to high levels of transparency such as Russia, China and Iran.

But such success is not reflected elsewhere, and it is here that we find the first failure of all of the diplomatic effort, aid programmes, and NGO campaigns focused on the resource curse issue, namely that all of them have been overwhelmingly focused on 'northern' (short-hand for North American and European), *publically-listed* multinational oil and mining companies. The popular dialogue around globalisation is replete with campaigns against the perfidy and general wickedness of such corporations. A large number of NGOs focus their lobbying efforts on said companies; most bilateral aid agencies and their employees uncritically accept this view of the world; and depressingly few foreign ministries and diplomats question it or understand it. If 'science diplomacy' is about understanding, analysing, translating and

[24]Though the exact rules on how this section of Dodd-Frank will be implemented are still very much in play. The SEC was initially sued by Oxfam America for delaying its publication of rules around how such reporting would work. When those rules were published in 2012 the SEC was then promptly sued by the American Petroleum Institute, providing a neat bookend to the inevitability that good regulation involves angering both sides equally. For more on this issue see http://www.revenuewatch.org/issues/dodd-frank.

explaining highly complex systems, and then bringing that analysis to the table, then this has been an epic fail by both the diplomats as well as those who seek to influence them.

The problem with the label 'Multinational Corporation' is that it fails to distinguish between firms that are publically listed and traded and those that are privately owned or state owned. The difference in behaviour between the three types of corporation is so vast that locating them under the same banner undermines the credibility of both diplomatic action and academic analysis. It is the premise of this chapter that publically-listed multinational corporations are most likely to be the targets of NGO advocacy campaigns and/or participants in international initiatives, even though they are in most cases minor players in the extraction and trading of resources; and that while they are not always clean, they are rarely the locus for the most egregiously corrupt behaviour in developing countries. In fact, when it comes to transparency and accountability publically-listed multinationals are invariably subject to more transparent reporting of financial results, greater scrutiny from stock market regulators, and heightened pressure from investors. Privately owned or state owned corporations are subject to few of these factors.

There is no doubt a good paper to be written on how publically-listed multinational corporations became the monster under the bed of globalisation and the supposed primary perpetrators of the resource curse. In the oil and gas sector it may well come down to a mistaken belief that those multinational companies that almost always have a dominant position in the retailing of oil and gas are also responsible for the extraction of those resources. On another level it may simply come down to lobbying laziness — that NGOs, development workers, and diplomats have showed little interest in understanding any corporation that you can't buy shares in or that doesn't have a nice office in a nice developed country capital city.

The omission of state-owned corporations from the oil and gas sector is a crucial one. Publically-listed multinational corporations may be who we buy from at the pump, but state-owned companies hold 85 percent of global oil reserves and account for 55 percent of global oil production.[25] In some cases these companies are directly involved in the production of oil and gas; in other

[25]The dominant state-owned players in global oil production are the national oil companies of Algeria, Iran, Iraq, Kuwait, Libya, Mexico, Nigeria, Qatar, Russia, Saudia Arabia, UAE, and Venezuela (EIA, 2012).

cases they are major shareholders in consortiums in which a multinational corporation will act as the operator of a number of fields and/or facilities.[26]

While many state-owned oil companies are prominent only in their own countries, the sheer amount of global oil production that they are responsible for leaves a very large gap in the coverage of many global initiatives and standards. Moreover, some of those companies are now becoming major multinational corporations in their own right; oil companies such as Norway's Statoil, Brazil's Petrobras, China's Sinopec, China National Offshore Oil Corporation (CNOOC) and Petrochina, and Malaysia's Petronas, all have substantial overseas exploration and production programmes.

Al Jazeera's fascinating documentary series, *The Secret of the Seven Sisters* takes an in-depth look at the history of the oil industry and in particular of the role of the so-called 'seven sisters' (BP, Shell, Exxon, Mobil, etc.) in the development of the industry. The documentary lingers extensively on the history of these companies from when they were state-owned companies themselves and were at times poorly disguised instruments of national foreign policy. It is this history that perhaps best explains the way that these corporations continue to be perceived. And while there is plenty to disagree with in the series, it concludes with the following observation:

> At the end of the 1960s, the Seven Sisters, the major oil companies, controlled 85 percent of the world's oil reserves. Today, they control just 10 percent... Nationalisation of oil reserves around the world has ushered in a new generation of oil companies all vying for a slice of the oil pie...These are the new Seven Sisters...Mainly state-owned, the new Seven Sisters [Saudi Aramco, Gazprom, Petrochina, NIOC, PDVSA, Petrobras, and Petronas] control a third of the world's oil and gas production, and more than a third of the world's reserves. The old Seven Sisters, by comparison, produce a tenth of the world's oil, and control only three percent of the reserves. The balance has shifted.[27]

---

[26]The Shell Petroleum Development Company in Nigeria, for example, is operated by Shell but has been majority owned by the state-owned Nigerian National Petroleum Corporation since 1974.

[27]Note: The percentages quoted here are different from the EIA ones presumably because the EIA figures look at all of the state-owned players, whereas the documentary's figures only look at the seven largest state-owned companies. While the documentary usefully covers this transition, the amount of time dedicated to the past behaviour of such companies undermines its final conclusion which is that there are new players on the block who are more significant and about whom we know very little. See Aljazeera (2013).

The second omission of privately owned (i.e. unlisted) oil and mining companies is also important. Some such companies are extremely significant — commodities trading firms Trafigura and Vitol had reportedly had revenues of US$122 billion and US$297 billion respectively in 2011[28] though as privately held companies any information they disclose is information they choose to disclose, rather than information that is required of them by regulators such as the SEC.[29]

At the other end of the spectrum the mining industry and to a less extent the oil and gas industry is replete with a number of small privately owned companies, which may be significant in only one or two countries, and may not be large enough to come to the sustained attention of NGOs, aid workers and diplomats. In the mining sector this is driven by the structure of the sector in which the vast majority of Greenfields exploration is done not by the big multinationals but by relatively small, sometimes innovative and risk-taking 'juniors'. What is certainly true is that once exploration moves to production there are still a significant number of small to medium sized private and unlisted companies who operate largely under the radar. At best these companies are able — free from the short-to-medium term focus of analysts and investors — to take a genuinely long-term approach to resource development. At worst the absence of scrutiny that comes with public listing can have a corrosive effect on governance in the countries in which they operate.[30]

---

[28]Very limited information on both of these firms is provided in 'corporate brochures' which are available at http://www.trafigura.com/site-information/download-corporate-brochure/corporate-brochure-en/ and http://www.vitol.com/brochures.html. It is noteworthy that is far easier to find detailed financial information about state-owned Chinese oil companies than it is for these two privately-owned commodities trading companies.

[29]The financial relationships between state-owned oil companies and highly secretive privately-owned commodities trading firms has finally become the subject of some extremely useful research by the Revenue Watch Institute and the Berne Declaration — see RevenueWatch (2013).

[30]One of my dominant experiences as I travelled in West Africa, the former Soviet states, and the Asia Pacific was that of constantly discovering many of these kinds of companies. They were very rarely publicly-listed anywhere, and the structure of their ownership often extremely difficult to determine. They were commonly headed by one or two prominent businessmen, and the consistent rumour would be that their greatest comparative advantage was not in geological or engineering expertise, but rather in their ability to 'interact' with government ministers and senior officials.

Someone Else's Problem

The next significant failure of diplomatic efforts with respect to resource curse issues is the fact that the focus has been overwhelmingly on producing companies and countries rather than on consumers. A generous interpretation of why this is would be that because the resource curse has affected mainly developing countries it has been a subject mainly examined by aid and development NGOs, bilateral and multilateral development agencies, and foreign ministries. The almost inevitable result of these kinds of organisations being the loci for action on the resource curse is that it has almost entirely been conceptualised, analysed and solved as a problem that is 'over there'; something to be solved either in developing countries or in the large corporations that people know to operate in those countries.

Those same NGOs, development agencies and foreign ministries are inevitably staffed by people who are disinterested in domestic policy. Their funders, whether they be private donors or finance ministries, often impose funding rules that require their administrative costs to be minimised and for the bulk of their income to be distributed in developing countries. If you're an aid organisation or a foreign ministry it's supposedly not a 'good look' to spend too much money at home. All of these factors combine to make aid and development organisations at best clumsy domestic advocates and players, and at worst actually unable to run programmes focused on changing behaviours in their own countries, even though the end point for the vast majority of oil and mineral resources are consumers in developed countries.

This is not to say that there has not been some effort put into changing how consumers in developed countries behave. Most prominently the Forest Stewardship Council and the Marine Stewardship Council have developed credible standards and trademarks around the sustainable management of forests/timber exports, and fisheries.[31] Each has developed a consumer trademark which companies adhering to their standards can then use when promoting their goods. To a lesser degree the Kimberley Process has attempted something similar in the diamond market, though neither Kimberley, the

[31] More information on these organisations is available at http://www.fsc.org and http://www.msc.org.

Forest Stewardship Council (FSC) nor the Marine Stewardship Council (MSC) have had the resources to carry out large-scale sustained marketing campaigns in consumer countries.

But such a focus on consumers has not flowed on into the oil and mineral markets. The majority of consumers know very little about origins of the goods that they consume, and those that do know something about them are very reluctant to sacrifice price for the consideration of externalities that impact on faraway people, countries or environments. The few who do take into consideration such issues tend to be focused on the environmental consequences of a good's production or producer, and pay little attention to the 'political externalities' (e.g., corruption, authoritarian government, and conflict) of the goods.

At the most absurd end of the spectrum, environmental organisations in developed countries that have allied themselves with those in the political spectrum likely to be distrustful of publicly listed multinational corporations, have managed to create a movement that will protest at the mine site or the oil well, but never at the steel smelter, the jewellery store, or the petrol station. This kind of action only reinforces the mental break between consumption and production, and allows citizens of developed countries to live in a fantasy world in which there are bad corporations but never bad consumers. Though as a counter-balance, the argument of 'you consume it therefore you're a hypocrite if you object to anything' is almost as blunt a rhetorical and analytical tool as that of ignoring consumption altogether.

The plea here is for some balance between the two positions, which would be a vast improvement from the current position of the role of consumers being ignored. The failure to achieve such a balance will make it all the more likely that oil and mineral production will increasingly be pushed out of developed countries and into developing countries where environmental regulatory standards will be lower or rarely enforced; where civil society organisations will be weaker; and where corruption and conflict is more likely.

Fine Clothes for the Wrong Emperor

Foreign ministries and diplomats, bilateral and multilateral aid agencies, NGOs and publicly-listed multinational oil and mining corporations have all put a significant amount of effort over the past decade into countering the resource curse. The issue remains a vital one. The revenues generated by these industries do and will continue to dwarf all aid and development funding globally. Accounted for and spent well they have the capacity to be positively transformative of a huge number of countries. Managed poorly and the benefits will accrue only to a generation or two of kleptocrats and/ or they will be so destabilising as to generate or sustain conflicts around the world.

The focus of most of these organisations, to force a metaphor, has been to dress the emperor in a set of fine clothes of good governance and accountability. The risk here of course is not so much that those clothes are not magnificent — the efforts of initiatives such as the EITI have had a positive impact — but rather that the emperor is a marginal figure, and that it's the courtiers in the shadows and the baying public outside that we should've been paying attention to all along.

Publicly-listed multinational corporations are a highly visible and convenient place to focus one's attention, and genuine scandals involving such companies occur regularly enough to perpetuate the belief that they are the most relevant players in this area. The irony, of course, is that it is often the fact that they are publicly-listed that generates the financial disclosures and regulatory scrutiny that allows people to identify misdeeds. Of far greater significance, however, are the multinational state owned or privately owned corporations that are often bigger players in the market; are less known to or understood by diplomats, aid workers or NGOs; and are often more complicit in corrupt behaviour.

The act of intellectual self-lobotomisation that consumers carry out on a daily basis — in which consciously or unconsciously we deny the link between what we consume and where and how it is produced — is perhaps one of the most crucial unresolved issues of globalisation. Those organisations involved in pursuing solutions to the resource curse have thus far proved themselves unwilling or unable to seek consumer-driven solutions. Some dismiss efforts at changing consumer behaviour as being too hard. Yet the defining premise of most aid and development organisations and their employees is that it is possible to have deep understanding of cultural, social, economic, historical, and political drivers of dozens of developing countries and to design interventions and programmes that will make them less poor.

Against this backdrop, changing consumer behaviour in our own countries looks like not only a simple task, but one which we personally have far greater understanding of and responsibility for.

Science diplomacy should, one would hope, be about an ability to integrate the understanding of highly complex systems into international negotiations and governance. And yet in many arenas it seems that diplomacy fails to focus on anything but the most visible participants in the ecosystem. Whether the reasons behind this are conspiratorial (knowing ignorance — deliberate reluctance to deal with an issue) or accidental (unknown ignorance — not doing anything about an issue because one is not even aware that it is an issue) is a discussion for another book. As is perhaps a concluding question, which is this — will diplomacy as it is commonly conceived ever be able to address such complex questions? Diplomacy is inherently prejudiced by the fact that it is a tool of nation states, and is largely exercised by small elites with an indifference or distain for domestic politics. This makes it a very poor instrument for dealing with multinational corporations of all varieties, let alone for influencing domestic consumers.

References

Al Jazeera (2013) 'The Secret of the Seven Sisters', *Al Jazeera: Special Series*, 26 April, [Online], Available: http://www.aljazeera.com/programmes/specialseries/2013/04/201344105231487582.html [20 May 2014].

Corruption Perception Index (CPI) (2011) 'Overview', *Transparency International*, [Online], Available: http://www.transparency.org/cpi2011.

Energy Information Administration (EIA) (2012) *Who are the major player supplying the world oil market?*, [Online], Available at: http://www.eia.gov/energy_in_brief/world_oil_market.cfm [20 May 2014].

Freedom House (2012) *Freedom in the world*, [Online], Available: http://www.freedomhouse.org/report-types/freedom-world.

International Monetary Fund (IMF) (2007) *Guide on Resource Revenue Transparency*, 15 May, [Online], Available: http://www.imf.org/external/np/pp/2007/eng/051507g.pdf.

International Monetary Fund (IMF) (2012) *World Economic Outlook Database*, April, [Online], Available: http://www.imf.org/external/pubs/ft/weo/2012/01/weodata/download.aspx.

Myers, K. (2005) 'Petroleum, Poverty and Security', *Chatham House Briefing Paper*, London: Chatham House, [Online], Available: http://www.chathamhouse.org/sites/default/files/public/Research/Africa/bppetroleum.pdf.

RevenueWatch (2012) *Bringing Transparency to National Oil Companies and 'Citizens' Oil'*, 23 April, [Online], Available: http://www.revenuewatch.org/news/press_releases/bringing-transparency-national-oil-companies-and-citizens-oil [20 May 2014].

Chapter 8

The Role of Science Communication in International Diplomacy

Joan Leach

The goal of this chapter is to lay out the problems and possibilities for science communication in international diplomacy. The context is Western democracy's drive for an instrument that makes science accessible for a majority of citizens. What is meant by science communication in this context varies rather widely, from science and technology promotion and marketing activities to science and technology education, and even programs to aid researchers with basic communication skills in their own and other languages. Also included in this cluster of activity are structures aimed at integrating science in the decision-making processes of contemporary democracies, better known as 'science in society' programs. Some of these may also take on a 'citizen science' flavour, whereby non-scientists are encouraged to help with data gathering or routine scientific work and thus enlarge the population of science-aware citizens (Irwin, 1995). For Western democracies the difficult question of public dissent about science and technologies has also meant that surveys of public attitudes to science have abounded and methods of producing consensus proposed. These actions are said to be guided toward improved science communication and public engagement with science.

What does the emergence of these activities tell us about the possible roles that science communication has in larger international moves to science diplomacy? What are the useful frameworks in which we can characterize this activity? What evidence is there (or could there be) that improved science communication helps with international diplomacy? This analysis will frame science communication in the contemporary discussion about soft power resources in international diplomacy. For science communication — in any of its formulations — to matter in diplomacy or international relations, one must at least entertain a notion of 'soft power' whereby persuasion and attraction, artefacts of communication, can play a role separate from actual or threatened force, retaliation, or payment (Nye, 2011).

In general, science communication should be an area that figures prominently in discussions of soft power, but frequently the communication aspect is assumed and not developed in analyses of the sciences' soft power resources. In short, there are abundant examples of scientists working across national borders and creating sustainable links among science communities as well as leveraging national and international resources. In instrumental terms, the tool they use to accomplish this is largely communication. In addition to scientists, however, science journalists and professional communicators also perform important communication roles that cross national boundaries and can be put to work for various diplomatic purposes. These activities have not gone un-noted among analysts and critics of science policy and international diplomacy (Stirling, 2010; Elyis and Tyfield, 2013). It is safe to say that contentious international issues can at least be articulated for further scrutiny by better communication among scientists and various international publics. Media, too, can play a role here in articulating and framing various issues that encourage further public interest.

Necessarily, then, 'science communication' becomes an umbrella term that covers direct expert-to-expert communication, popularization through public and private media outlets, publicisation through various means of advocacy, and even dialogue between advocates and dissenters of science and technology. With such a broad set of definitions, it is useful to set out some distinctions that relate to previous work on science diplomacy and international relations in general. In this discussion, I am going to use the following arenas of science communication for clarity while admitting up front that highlighting these arenas emphasizes the instrumental uses of communication at the peril of more nuanced political analysis, to which I will return in the conclusion. First is the rather obvious arena of 'professional science communication' which refers to communication about science and technology by researchers for the purposes of furthering scientific and technical research. In short, it is communication by researchers for researchers. Second, science communication can also include science popularization (including science promotion and marketing as well as science journalism). This communication is directed at multiple audiences and can be produced by researchers or professional mediators such as journalists or public relations professionals. It may be that a better term for it is 'publicisation', in the sense that these communication activities make science public in various ways (Raman, 2014). The point of it is to disseminate the findings, methods, or nature of research to audiences other than the scientific community. The purposes of this activity could be broadly democratic, to encourage engagement or debate about the science, report on results of public funding, or educate an interested group in the nuances of a particular research field. Purposes, however, can also include raising the readership numbers of a media outlet, pursuing public relations goals for an institution and even chasing increased funding. Finally, there is the arena of science communication policy whereby national and international institutions attempt to encourage, discourage, guide and constrain scientific communication at both national and international levels for a range of purposes. This final arena of science communication, in national policy, has the potential to affect international development and international relations in important ways. Encouraging or constraining science communication can effectively create conversations or gag potential participants.

To date, discussions of science diplomacy have been broken down into 'diplomacy for science', aimed at facilitating international science cooperation especially in the case of large-scale scientific projects, 'science in diplomacy', which focuses on informing policy objectives with scientific advice, and 'science for diplomacy' which uses scientific cooperation to improve international relations between countries (Lord and Turekian, 2007). The science communication arenas outlined above map onto these various approaches to science diplomacy in myriad ways. The chart below illustrates the ways in which various forms of science communication intersect Lord and Turekian's (2007) breakdown of the forms of science diplomacy.

Science Communication Supporting Diplomacy for Science

When researchers can articulately argue for international cooperation in support of large-scale projects, their very ability to communicate in this context, somewhat removed from the technical communication of everyday science, is a special resource. Additionally, when a nation has a wealth of science popularization that can effectively communicate the nature, goals, and results

Table 1. Types of science diplomacy and science communication activities.

	Professional Science Communication	Popularization or Publicization of Science	Science Communication Policy
Diplomacy for Science Example: SKA (Square Kilometre Array[a]) in Australia, New Zealand, and South Africa.	Researchers communicating to establish large-scale, international cooperation in science	Public relations and journalism activities raising awareness of outcomes of large-scale international projects	National attempts to give researchers skills in organizational communication, negotiation, and intellectual property issues
Science in Diplomacy Examples: IPCC (Intergovernmental Panel on Climate Change), but also various Parliamentary and executive officers and committees giving advice to governments.	Initiatives by research bodies to communicate the potential of research to solve policy problems, to open dialogue, and give advice about the regulation of science and technology.	Targeting policy-makers as a key audience for research results and outcomes	National attempts to encourage researchers to communicate with policy makers and embed research in governmental processes
Science for Diplomacy Examples: AAAS Centee for Science Diplomacy, and the Danish Board of Technology.	Researchers communicate with collaborators across national divides and despite restrictions (i.e., US-Cuba collaboration)	Popularization encouraging high levels of general scientific literacy, awareness, and dialogue about science and technology (i.e., PUS, PEST, Public Engagement initiatives)	National encouragement and support for international research through communication skills, cultural programs, and language programs to increase capacity for international collaboration as well as facilitate international dialogue about contested science and technology.

[a] For the official website of SKA, see http://www.ska.gov.au.

of technical research to audiences within and outside that nation, this, too, can be a substantial persuasive resource. The recent example of the Square Kilometre Array (SKA) provides insight into how science communication can function as a national resource for cooperation and competition. The SKA will be the world's largest radio-telescope. In 2012, Australia (including a component in New Zealand) and South Africa (including support from nine other African nations) were chosen as sites for the dish antennas that are the core of SKA. This result, itself a compromise between siting the SKA wholly in South Africa or Australia, came at the end of a massive scientific as well as science communication campaign by Australia/NZ and South Africa to demonstrate to both a scientific audience and an attendant public that each nation deserved to host this big science project.

One of Australia's leading science broadcasters, 'Dr Karl', was enlisted to promote the SKA to popular audiences in Australia and raise public awareness. Given that he has his own science-themed show on the national broadcaster, the ABC, he is an example of a celebrity science communicator who has more reach to more audiences than general journalists (News.com.au, 2012):

> ...using his energetic communication style, Dr Karl told Australians about the size of the SKA and its huge potential for discovery. He also explained how the whole world benefits from mega-science projects of this nature through education and technical spin-offs.

Journalists have also popularized the SKA internationally, suggesting that its scientific merit is coupled with economic advantage for the hosting nations and region (Cooksen, 2010). The public nature of a big-scale project also incites arguments about which global region is 'deserving' of hosting such a large-scale scientific project. The Australian bid underscored the high levels of scientific and technical literacy in Australia. South Africa's science minister responded to the existing public support for SKA in Australia/NZ with a bid to boost scientific capacity in Africa: 'We see this as an opportunity to inspire our young people through astronomy to take up careers in science' (Cooksen, 2010). Thus, as far as diplomacy for science goes, science communication contributes soft power resources to individual nations as they cooperate and compete for large-scale projects. Promoters are enlisted, journalists engage audiences with the back story, and researchers who are talented communicators lobby publics and governments for support.

To describe this in terms of 'soft power' instead of 'business as usual' is to point out several features about the field of science communication. First, instead of being something scientists just 'do', science communication efforts, both by researchers and by public relations officers, and extending to media, are coordinated and organized, especially in the case of large-scale international projects. Second, the storytelling that appeals to publics is increasingly being used to leverage support for science and technology. Third, designating these activities under the banner of 'soft power' underscores the involvement of government resources in these activities. This is not new. The science journalist Daniel Greenberg (1968) has described the increasing organization of state resources around science, for diplomatic and other ends, since the 1960s and historian Roy MacLeod (1997) has traced advocacy in science policy over several centuries. So, the emphasis on soft power sheds light on the advocacy role of science policy and the concerted efforts that scientific organizations have made to create communication channels that they can use to promote scientific projects.

Communicating Science for Diplomacy

Additionally, there is a growing trend toward national legislation of science communication, clarifying the role that researchers have in relation to state interests. This has been most marked in Australia, Europe, the UK, and India, but China and Japan have also explored national statements and institutions that will guide or be responsible for science communication. Such policies indicate that science communication is starting to be seen as a soft power resource that needs careful development as well as some state control if it is to be effectively tapped as a resource in political contexts. This highlights the rather uneasy relationship between science communication as 'science promotion' and science communication as a public service oriented activity by journalists, educators, and researchers with very different motivations for communicating. In Australia, for example, the commonwealth government issued 'charters for science communication' with government sponsored institutions (Australian Commonwealth Charter). These charters were to ensure that researchers saw communication as part of their professional duty and responsibility. However, these charters also underscored that the nature of some research might be sensitive to national interests and communication about it should be restricted. Canada has come under fire in similar circumstances in an editorial in the journal *Nature* (Anon, 2012) for 'muzzling its scientists' and 'prioritizing message control and showing little understanding of the importance of the free flow of scientific knowledge'. Such situations show the difficulties, in terms of science diplomacy, of coordinating a 'national' (let alone a global) approach to controversial areas such as climate change, bio-engineering, nanotechnology, and the like. While science promoters can be given a unified message to present about government-sponsored research, researchers might have different motives and ends for their research and communicate that quite effectively. Of course, journalists can also work to subvert the messaging of governments in science promotion, pointing to risks and hazards, complications, hype, and other negative or ambiguous aspects of science and technology. Academics and researchers across fields are becoming more visible in national and international debates about the communication of benefits and disadvantages of scientific research and technical solutions to policy problems.

From the point of view of researchers in science, the issue seems to be less about the efforts governments make to constrain their communication, but rather what access researchers have to policy-makers in order to influence science policy more broadly. One example of an initiative designed to bring researchers into conversation with policy-makers has been 'science in parliament', attempted in both the UK and Australia (SIP, 2012). As few politicians have scientific backgrounds, the efforts seem to be based on advocacy for science in general, but also for increased awareness of specific scientific projects worthy of broader political recognition. There is also no shortage of public relations campaigning on behalf of scientific institutions to make policy-makers aware of research and funding needs (see for example, Reiff (2003) and Cotgreave (2003)). Many of these efforts are spearheaded by former researchers retooling their careers into public relations in an effort to increase funding to their areas of previous research interest. However, there can be no doubt that these efforts have established lines of communication between policy and science, albeit in ways that will favour already well-funded areas.

Much rarer have been attempts by governments to embed researchers into the policy process. One pattern is to have offices of science and technology as part of the budget office (in the US it is part of the White House), or offices of the chief scientist (Australia and New Zealand) or the Parliamentary Office of Science and Technology (UK) where expert scientific and technical advice can be channelled to government. Rarer still is the effort made to integrate scientific and technical concerns, social concerns, and governance in such a way that increases national and international awareness and debates about the governance of science and technology. One celebrated example is the Danish Board of Technology, an independent body that encourages pubic debates about science and technology and forwards the results of such discussion to the Danish parliament. What is crucial about this approach is that it does not assume consensus on complex issues where there is no evidence of consensus and it can 'open up' discussion instead of closing it down. This rare but important example suggests that one role for the scientific diplomat may be to open discussion on relevant scientific issues rather than advocate for any one solution. This is less wishful thinking than it would have been in the past. For example, a 2009 meta-analysis in the Netherlands reviewed the evidence base for the value of public engagement by scientists (Benneworth, 2009). They found that there is a small but marked shift toward open public engagement as opposed to direct science advocacy in Western democracies. But, they also found that in every national survey, respondents did not trust governance structures to do anything that mattered with the opinions they shared during engagement sessions. That is, there was a widespread feeling that engagement did not really mean anything for changing the outcomes of scientific controversy, debate, or regulation. This points to a challenging context for science communication that encourages diplomatic relations. Expectations seem to be growing that public engagement with science should be a communication process that is open and in a venue where the outcome is not a foregone conclusion. Coupled with awareness that engagement processes are not hooked up to governance practices, diplomacy faces a challenge on two fronts. First, science diplomats need to communicate for engagement, not advocacy. But, even if this can be achieved, diplomatic efforts may be greeted with some scepticism unless there is evidence that engagement actually means something for visible governance.

Science Communication for Diplomacy

In his discussions of 'soft power', Joseph Nye (2011: 226) highlighted the results of the Pew Global Attitudes Project in 2011 indicating that, despite growing negative attitudes about America in general, every country surveyed indicated that a majority of interviewees 'admired the US for technical and scientific advances'. Nye includes this marker in his discussion of 'culture' as a soft power resource and science might be the area which best makes the case for Nye's discussion of soft power, and more recently, 'smart' power (Zewail, 2010). Such discussions are not limited to the United States. Flink and Schreiterer (2010) characterize three goals that nations have in promoting international scientific cooperation, two of which are heavily reliant on science communication. First, they discuss access-driven initiatives whereby cooperation between two nations increases at least one of the nation's ability to use the other's scientific infrastructure. This might be for the education of researchers (Universities in the US and UK have long enjoyed top status as destinations of choice for scientific research training) or for access to specialized equipment too rare or expensive for one nation to support (i.e., the Large Hadron Collider or the SKA). But while communication interactions in these scenarios quite possibly impact science diplomacy, it is science promotion and influence-seeking that Flink and Schreiterer suggest as goals that have profound soft power implications for science diplomacy. In the case of science promotion, it is an oft-repeated claim that science is unique as a mode of knowledge production in that it can stand outside of politics (Brown, 2009). This claim enables collaborative research and communication in scenarios where political structures would be a deterrent (e.g. during the Cold War between European and Russian scientists). More recently, science has been touted as the basis for 'the deepest link' between the US and Muslim World (Anthis, 2009). Again, the key idea is that 'the universality of basic science' provides a baseline for conversations that stand outside politics. It is also the case that science promotion need not *prima facie* appear as national promotion. A scientific or technical success can be claimed for humanity, even as the achievement is heralded as a national success. Neil Armstrong's assertion that his first steps on the moon were '...one giant leap for mankind...' even as he placed an American flag on its surface is the paradigm case of this dual rhetoric of national achievement and human progress.

Recent moves in the US to send science envoys around the world rests on this idea that science provides a neutral political space for discussing common or pressing problems. But, as Sheila Jasanoff has argued, these envoys might be in peril if they assume that science itself offers an obvious basis for mutual understanding and transparent communication (Jasanoff, 2009). Jasanoff worries that there are serious misconceptions that could hinder the success of science envoys. Among these, she identifies core problems of science communication. The first for communicators is 'who is your audience?' One worry is that science envoys assume that 'more science means more progress' for everyone and that the science, which is their reference point, is American science. This could result in envoys presenting answers to scientific questions that they bring with them from the US Jasanoff suggests that science envoys need to first ask, 'whose problems are we solving?' This question is not only relevant to the content of scientific problems, but also to the form.

It is increasingly difficult for those outside the scientific discussion to frame questions that make sense within scientific discussions. It is all too easy to see non-scientists concerns framed as outside the discussion. Publics are not, after all, communicating in a scientific way, but rather a way that is seen as 'from the outside' or even 'irrational'. Thus, scientific envoys run the risk of only talking to other scientific envoys and not engaging deeply with the concerns (some possibly resolved or addressed by science and technology) of those outside the scientific establishment. In a damning historical parallel, the 'Republic of Letters' which was the enactment of scientific communication in the Enlightenment did not necessarily address the problems of starvation and disease in the actual monarchies and oligarchies in which learned letters circulated (Stirling, 2010). Jasanoff signals that we should not expect scientific envoys communicating purely in scientific language to be successful in demonstrating commonalities outside science, and warns against assuming that science diplomacy will 'promote cross-cultural understanding'. Behind this, she identifies another common assumption, 'that scientists everywhere share a commitment to disinterested inquiry and communal problem solving, and hence can open up channels of communication when others fail' (Jasanoff, 2009). While this may work for circumscribed problems in contexts where there is prior agreement about communication processes and goals, there is little reason to expect that scientific expertise will succeed when other diplomatic channels have failed. This will be especially true when problems are complex, goals are unclear, and motives differ for engaging in communication in the first place.

Conclusion — Rethinking Science Communication for International Diplomacy

On a positive note, science diplomacy can encourage cooperation and 'international scientific cooperation comes to be seen as an effective agent to manage conflicts, improve global understanding, lay grounds of mutual respect and contribute to capacity-building in deprived world regions' (Flink and Schreiterer, 2010). Promoting or marketing national achievements in science and technology can contribute to 'soft power' resources that have the potential to produce diplomatic outcomes. It is also the case that formal science communication policies and sustained national promotion campaigns featuring science and technology are on the rise globally. For analysts, this is an interesting development in the story of how science is integrated (or not) in culture. However, this chapter's cautionary tale is that like science, communication is not an unproblematic 'good'. Success in communication, whether by scientists or professional communicators, is not a foregone conclusion.

The issue of success in science communication is also complex. Researchers' success is not necessarily diplomacy's success; it is possible that complex problems understood through a scientific lens could lead to the rejection of desired diplomatic outcomes. The relationship between scientific motives, diplomatic motives, and democratic motives is also fraught. While communication viewed instrumentally can be in the service of any of these motives, thinking about science communication as a fundamental precursor to mutual engagement of complex problems where the outcome is not predetermined may eventually prove a more reasonable stance for international science diplomacy. The consideration of 'science communication' in this discussion has been fairly simple and instrumentalised in order to set out some categories for analysis as well as eventual deconstruction. As Mark Brown (2009) argues, communication in and about science has itself been a topic of political consideration in the social history of science. Communication has long been held to stand for openness rather than secrecy and for democracy rather than elitism, and the term itself has often been open to jingo-ism (see Dewey (1934) for a famous example; and Peters (1999) for an overview of the idea of communication). The ideal of communication as openness coupled with a value-free ideal of science (Douglas, 2009) makes for a powerful rhetoric in favour of scientists pursuing diplomacy with the tool of science communication. Putting science communication into international diplomacy also makes sense in the context of soft power approaches to international diplomacy.

However, in canvassing the role of science communication in the three approaches to science diplomacy defined by Lord and Turekian (2007), the instrumental approach to science communication proves inadequate to understanding both the reality and the potential of science communication in science diplomacy. In Lord and Turekian's first category, *diplomacy for science*, science communication in instrumental terms is how science is promoted. Structurally, what stands out in this brief analysis of science communication in diplomacy for science is the remarkable convergence of communication strategies among scientific researchers, journalists, and governments in support of science. What has been called the 'soft power of science' now seems to be mobilized by the 'soft power of science communication'. Science communication is not just an instrument of advocacy; it is a term that marks the strategic orientation of science to publics and audiences globally.

In Lord and Turekian's second category, *science in diplomacy*, science communication in instrumental terms is what scientists need to learn in order to effectively work with diplomatic and political structures. But, again, the instrumental view of communication falls short. Science communication is increasingly not being seen as a one-way dissemination of information or even a two-way conversation. Rather, science communication is being reformulated as a process of engagement where publics and scientists can air their views but with an expectation that those views will have an impact on the governance of science. So, rather than science communication being the tool that can inject science into diplomacy, science communication becomes the occasion and venue for engagement for both publics and science diplomats.

Thirdly, Lord and Turekian introduce the notion of *science for diplomacy*. Under this description, science communication is the instrument for openness and value neutrality, potentially working through diplomatic channels that have failed in the past. Again, as Jasanoff warns, it is dangerous to assume the value neutrality of science as well as the strength of one's own stance in communication with other cultures. Here, science communication can profitably help the potential diplomat with the fundamental questions of communication — to whom do I speak and why? The 'how' follows from interrogating those questions. For science communication, the question of 'why' needs to be substantially broadened for the science diplomat. The science diplomat need not be public advocate for any particular scientific outcome; rather the diplomat needs to be an advocate for science communication itself — an advocate of the open sort of engagement that will need, on occasion, to acknowledge dissent and difference.

Finally, it is tempting to add to Lord and Turekian's typology. Missing from their 2007 discussion is diplomacy *in* science. Much has recently been written about sciences' social license to operate (Raman, 2014) which is tacitly given by publics to governments and scientific researchers. It is based on a form of social trust and it has proven to be a fragile and revocable license. Western democracies, suggests Wilsdon, have been rethinking the social license for science, wondering if the promises of new technologies and knowledge are worth the environmental and social perils of their antecedents. Jasanoff has written that scientists themselves need to embrace new technologies — technologies of humility — to begin formulating a new social license. Science communication of the open, engaged sort, could be such a technology of humility, especially if researchers would see themselves engaged in diplomacy in science itself. This implies that that their achievements are acknowledged, but also that the frameworks in which scientists produce knowledge and technology are given by increasingly global publics. This view broadens both the notion of science diplomacy and that of science communication as a form of engagement. Science diplomacy is not just the job of an appointed science envoy, but the work of every researcher.

Cited References

Anonymous (2012) 'Frozen Out', *Nature,* 483: 6. DOI: 10.1038/483006a.

Anthis, N. (2009) 'A Universal Truth', *SeedMagazine* 17 September, [Online], Available: http://seedmagazine.com/content/article/a_universal_truth/.

Benneworth, P. (2009) 'The Challenges for 21st Century Science: A Review of the Evidence Base Surrounding the Value of Public Engagement by Scientists', Working paper, *Centre for Higher Education Policy Studies (CHEPS)*, [Online], Available: http://www.britishscienceassociation.org/NR/rdonlyres/C7CC70B3-1A6E-485D-B3C8-3E2BCC87653F/0/BenneworthFINAL.pdf [10 May 2014].

Brown, M. (2009) *Science in Democracy*. Princeton: MIT Press.

Cooksen, C. (2010) 'Race to Host the Most Expensive Telescope', *Financial Times Magazine*, 15 October, [Online], Available: http://www.ft.com/cms/s/2/ed627170-d66b-11df-81f0-00144feabdc0.html#axzz31hD2ZjW9.

Cotgreave, P. (2003) *Science for Survival: Scientific Research and the Public Interest.* London: British Library.

Dewey, J. (1934) *Art as Experience*, New York: Minton, Balch & Company.

Douglas, H. (2009) *Science, Policy, and the Value-Free Ideal.* Pittsburgh: University of Pittsburgh Press.

Elyis, A. and Tyfield, D. (2013) 'Citizens and Science in a Greener China', *The Guardian*, 16 October, [Online], Available: http://www.theguardian.com/science/political-science/2013/oct/16/citizens-science-greener-china.

Flink, T. and Schreiterer, U. (2010) 'Science diplomacy at the intersection of S&T policies and foreign affairs: Toward a typology of national approaches', *Science and Public Policy*, 37(9): 665–677.

Greenberg, D. (1968) *The Politics of Pure Science*. New York: New American Library.

Irwin, A. (1995) *Citizen Science*. London: Routledge.

Jasanoff, S. (2009) 'Lessons for Science Envoys', *SeedMagazine*, 17 September, [Online], Available: http://seedmagazine.com/content/article/lessons_for_science_envoys/.

Lord, K.M. and Turekin, V.C. (2007) 'Time for a New Era of Science Diplomacy', Science, 315(5813): 769–770.

MacLeod, R. (1997) 'Science and Democracy: Historical Reflections on Present Discontents', *Minerva*, 35(4): 1–16.

News.com.au (2012) 'Dr Karl fights for Australia to host massive Square Kilometre Array telescope', 18 January, [Online], Available: http://www.news.com.au/technology/crunch-time-to-win-outer-space-bid/story-e6frfro0-1226247007106.

Nye, J. (2011) *Soft Power*, 2nd edition, New York: Public Affairs.

Peters, J.D. (1999) *Speaking into the Air: A History of the Idea of Communication.* Chicago: University of Chicago Press.

Raman, S. and Mohr A. (2014) 'A Social License for Science: Capturing the Public or Co-constructing Research?' *Social Epistemology*, forthcoming.

Reif, H. (2003) 'Communicating Science to Policy-Makers and Policy to Scientists', 7 February, *Science Careers,* [Online], http://sciencecareers.sciencemag.org/

career_magazine/previous_issues/articles/2003_02_07/nodoi.1298499335 1735824631.

Science in Parliament (SIP) (2012) Science in Parliament, *Office of the Queensland Chief Scientist*, [Online], Available: http://www.chiefscientist.qld.gov.au/science-in-parliament.aspx.

Stirling, A. (2010) 'Keep it Complex', *Nature*, 468: 1029–1031. DOI: 10.1038/4681029a.

Zewail, A. (2010) 'The Soft Power of Science', *The American Interest*, 1 July, [Online], Available: http://www.the-american-interest.com/articles/2010/07/01/the-soft-power-of-science/.

Additional References

Australia. Department of Industry (n.d.) *Public Research Agency Charter with the Australian Nuclear Science and Technology Organisation*, Available: http://www.industry.gov.au/science/Documents/CharterANSTO.pdf [10 April 2013].

Jasanoff, S. (2007) Technologies of Humility, *Nature*, 450: 33. DOI: 10.1038/450033a.

Chapter 9

Science, Technology and WikiLeaks 'Cablegate': Implications for Diplomacy and International Relations

Daryl Copeland

The role and place of science and technology (S&T) in international affairs — and science diplomacy (SD) in particular — has attracted renewed interest in recent years. Given the nature of the primary threats and challenges which currently imperil life on our small planet — climate change, diminishing biodiversity, pandemic disease, resource scarcity, and environmental collapse, to name a few — this is perhaps unsurprising. And it is certainly welcome. Absent a better understanding of the S&T drivers central to each of these issues, and the development of a more effective capacity to manage S&T files, improved performance is unlikely.

This paper evaluates the unacknowledged significance of science, and science diplomacy in today's world. But it is equally concerned with technology, and in particular with the WikiLeaks 'Cablegate' episode as a case study of the impact of digital communications technology.

To frame and situate that discussion in the context of this volume requires a degree of analytical breadth. To get at the key connections, I have divided the paper into two sections. The first examines what I consider to be the essential issues related to the study of international science and technology as it relates to diplomacy and foreign policy. The second assesses the implications of 'Cablegate'.

Part I: Setting the Scene — What in the World are We Talking About?

Science is an evidence-based form of knowledge acquisition. It is founded upon empirical methods of experimentation and the repeated verification of results. Neither inherently political nor ideological, science is a type of universal language, a vector of transnational communications which poses fundamental questions about the nature of things. Science is typically bottom-up in origin, long term in orientation and collaborative by design. The findings of most scientific enquiry become part of the public realm. Crucially, science proceeds from the assumption that all events are caused, and all causes can — eventually — be determined. Problems, therefore, can be remedied. Poverty and suffering do not constitute necessary elements of the human condition. Adversity can be rolled back through the creation of new knowledge — to prevent and cure disease, discover alternative energy sources, invent new materials, and so forth. Science enlarges our understanding of the world and encourages broadly-based development.

Science also plays an important role in the formation and conditioning of intellectual culture and national values. In its scope and methodology, science helps to inform current analysis and to educate enquiring minds. The scientific ethos of objective experimentation through trial and error has broad appeal: it promotes merit (through peer review); openness (through publication); and civic values and citizen empowerment (through the encouragement of respect for diverse perspectives). In short, science advances learning in a transparent, participatory, and inclusive manner. It represents a cornerstone of humanity's progress. That may be why in public opinion survey research reported nationally in New Zealand, 20 June 2011, scientists were identified as the most trusted people in the country, and science as the most respected profession (TVNZ, 2011).

Technology, in contrast to science, is applied knowledge which mediates our interaction with the world. New applications, either intellectual or practical and especially those with transformative qualities, are referred to as *innovations*. The relationship of technology and innovation to science is not, as is widely believed, always linear.

Because technology touches more directly and immediately upon government and private sector interests, its development is often top-down, short term, competitive, and demand-driven. New technology may or may not be related to a specific program of scientific investigation. Moreover, as the possession and use of technology can confer advantage — a new weapon, a faster chip — the latest technological innovations are often licensed, sold, used to gain bargaining points, or otherwise protected as private goods. A tool in the hands of man, technology is related more closely than science to the possession and use of power, which is the capacity to achieve specified outcomes. Technology, therefore, tends to be regarded and used as an instrument of international policy, and in that context the deployment of technological innovations is often disruptive.

Diplomacy is a non-violent approach to the management of international relations characterized by dialogue, negotiation and compromise and featuring knowledge-based problem solving and complex balancing. Diplomats pursue and deliver international policy objectives on behalf of governments, and it is that connection to the state which sets diplomatic practice apart from the international lobbying, advocacy and public relations activities engaged in by business and civil society actors.

Science diplomacy (SD) combines the scientific method of knowledge production with international political agency. It is a crucial, if under-utilized specialty within the diplomatic constellation which can be used both to address global issues and to showcase national S&T capacity.[1] In this regard SD is a significant generator of *soft power* (Nye, 2004), that potent form of attraction that harnesses national image, reputation, and brand. More broadly, science diplomacy is an effective emissary of essential values such as evidence-based learning, openness and sharing. In its potential to address many of the planet's most urgent challenges, such as management of the global commons, public health and ecosystem collapse, SD is indispensible. As was so often the case during the Cold War, by using neutral, non-ideological language SD can be used to overcome, or at least mitigate

---

[1] A useful synopsis is offered in *New Frontiers in Science Diplomacy* (The Royal Society, 2010). This publication sets out three distinct activity areas within science diplomacy: informing foreign policy objectives with scientific advice (science in diplomacy); facilitating international science cooperation (diplomacy for science); using science cooperation to improve international relations between countries (science for diplomacy).

international political differences when regular diplomatic channels are strained or blocked.[2]

By virtue of science diplomacy's direct relationship to government interests and objectives, it differs from international scientific co-operation, which is sometimes commercially oriented and often occurs without direct state participation. International scientific co-operation is typically a win-win proposition, with private sector or civil society partners collaborating to produce, for example, better medications, cleaner water, improved hygiene, or more disease-resistant crops. All parties reap the rewards. Science diplomacy is also founded upon mutuality and common cause, but because national interests and the state are always implicated, motives may diverge and the outcomes may be asymmetrical, particularly if there are negotiations involved. A whole constellation of international scientific programs and exchanges undertaken during the second half of the last century come to mind by way of illustration, as do contemporary international discussions on issues such as the terms and conditions of resource access or environmental protection.

It must also be stressed that not all science diplomacy is devoted to the achievement of pacific ends. Covert collaboration involving, variously, Pakistan, Iran, China, North Korea, and Libya on nuclear-explosive and missile-propulsion technologies is an illustrative case in point.

But ... back to basics, to the *idea* of science itself. In a contested and competitive world of voodoo economics, bundled derivatives, radical politics and religious extremism, science proceeds from the assumption that misery is not fated: because all events are caused, all problems — eventually — can be solved.

At its best, science might be seen to represent the closest thing we have to universality, perhaps even truth. In the roiling realm of international relations, it merits considerably more attention than it has recently been accorded. Indeed, from all of the above it might be inferred that science, technology and diplomacy are likely to share much space in international

[2]This characteristic helps to explain the current focus within US foreign policy on expanding science diplomacy with the Arab and Islamic worlds, and it aptly illustrates the use of *science for diplomacy* (Lord and Turekin, 2007). An extraordinary, but all too rare multilateral example is the SESAME Synchrotron project in Jordan (Smith, 2012).

relations. SD, which brings together scientific know-how and diplomatic networks, should in particular occupy a central role in bridging the digital divide and addressing the essential issues of underdevelopment and insecurity.

If only...

Two Solitudes: Science and Technology; Diplomacy and International Policy

Science is a global enterprise, and is directly relevant to finding solutions to some of the world's most pressing problems. However, there exists a fundamental difficulty: S&T issues are largely alien to, and almost invisible within most international policy (IP) institutions. S&T, on the one hand, and IP, on the other, are effectively two solitudes, existing in separate floating worlds which rarely intersect.

When diplomats or politicians speak of IP, you rarely hear anything about S&T. Similarly, when scientists get together to discuss their work, it is rarely in the context of diplomacy or international policy. Indeed, most scientists cherish their independence from politics and government, while most politicians and diplomats are unfamiliar with S&T. The skill sets, activities, time frames and orientations of the two groups differ markedly.

> It must be asked: how many diplomats are scientists?
> How many scientists are diplomats?
> How often do scientists and diplomats mix?

Foreign ministries, development agencies, and indeed most multilateral organizations are without the scientific expertise, technological savvy, cultural pre-disposition, or research and development R&D network access and cross-cutting linkages required to understand and manage S&T issues effectively.

Add all of this up, and a rather disturbing picture emerges. It is something akin to a 'triple whammy'.

In mainstream popular culture: diplomacy is seen as irrelevant and ineffective; international policy is viewed as esoteric and exotic, and; science is perceived as complex and impenetrable.

Raise any one of these subjects separately and most people's eyes glaze over. Put all three together, and you have a combination capable of stopping just about any dinner party conversation in its tracks.

Even if the public environment was more solicitous, and scientists, politicians, diplomats and foreign ministries more favorably disposed and better equipped, major hurdles would remain. Public and private sector, NGO and university perspectives and interests are not always complimentary as regards S&T, R&D and innovation. Often they are contradictory or competitive.

Consider, for instance:

- the preponderance of private sector control over essential S&T intellectual property (patents and copyrights limit spread of innovation and transfer of technology);
- the influence of what President Eisenhower described as the Military Industrial Complex over funding priorities and research agendas (most governments are still spending more on defence research than on health research);
- the militarization of international policy more generally — defence departments have been accorded the lion's share of IP resources, while diplomacy and development assistance have been sidelined and marginalized, resulting in serious misallocations and distortions, especially at a time of resource reductions.

These observations provide some idea of the scope and dimensions of the challenge.

If this is to change, and in order to examine the remedial possibilities, political leaders and senior officials must be critically aware of the dynamic inter-relationships among principal actors and the key questions and issues at play.

Unfortunately, most are not. It is not just that the dots are not joined up. In most cases, there are no dots. These matters are not on the political map.[3]

[3]There are, of course, exceptions, and the practice of science diplomacy is more advanced in some countries — including, for example, the US, UK, Switzerland — than is the case in others, including Canada.

Improving Performance: What Can be Done?

At the top level of analysis, governments will require a grand strategy with a major S&T component.[4] Significant human resource, budgetary and institutional change will also be necessary.

Within the foreign ministry:

- Develop an International S&T policy and action plan, focussing on points where national capacity, domestic interests and global needs intersect (Infrastructure development? Wireless telephony? Public health?);
- Elevate science diplomacy and the tracking and evaluation of international S&T issues to priority status; establish this issue are as a specialty within the foreign service;
- Create the position of International S&T Advisor to the Deputy Head to insure that foreign policy development is informed by science ("science in diplomacy");
- Construct a bureau, under a Director General for International S&T, which would straddle trade and political sides of the organization and include a robust policy development capacity;
- Intensify ties with development agencies, science-based and industry departments, research councils, universities, think tanks, and NGOs; minimize legal and regulatory impediments to international collaboration;
- Increase capacity by bringing outside in and turning inside out through secondments, exchanges, internships and strategically targeted placements;
- Revamp the curriculum of diplomatic training academies; enlarge and S&T training and professional development opportunities.

At the national level:

- Rescind any extraordinary controls on scientific and diplomatic communications;
- Create, where it does not exist, the position of Science Advisor to the Head of government, and house the function at the executive or central agency level;

[4] Grand strategy is often overlooked as a component of international policy. For an assessment of the US example, see Zaharna (2010).

- Consider the establishment of an umbrella-style Ministry of State for Science and Technology as a standalone entity, with a mandate to improve linkages between those with the S&T knowledge and experience, and those responsible for public policy, administration and management.[5]

Internationally:

- Establish partnerships in education and training programs to upgrade the stock of skills and human capital and increase capacity;
- Focus on emerging technologies and platforms (for instance, bio- and nanotechnologies; information and communications technologies; genomics; new materials);
- Develop and implement strategies to strengthen national innovation systems, networks and infrastructure; nurture an enabling environment;
- Connect advantageously with the private sector in developing new approaches to technology transfer through foreign direct investment (FDI) and a focus on small and medium-sized enterprises (SMEs), private philanthropy and venture capital;
- Leverage global value chains and encourage investment partnerships;
- Build indigenous capacity and implement measures to foster institutional linkages (public/private; national/international) and networks; create new forums and spaces for dialogue and cooperation;
- Address the most pressing issues of development (urbanization, environment, water, food) through the creation of new mechanisms at the points of intersection between local needs, global S&T capability and commercial opportunities;
- Develop practical benchmarks and indicators to measure and evaluate progress.

This is an extensive 'to do' list, but it is really just a start. The magnitude of work to be done cannot be underestimated or trivialized, but such is the task at hand.

Much of the analysis contained in this volume is devoted an examination of various elements of 'science diplomacy' and accordingly will draw

[5]The government of New Zealand established a new Ministry of Science and Innovation in early 2011. It was folded into a larger department a year later.

substantially on the themes and issues set out above. The balance of this paper, however, will diverge from that path.

Part II: Freedom of Information: Everywhere and All the Time?

Not unlike the characterization of 'science in diplomacy' popularized by my colleague Vaughan Turekian, in previous writings I have attempted to assess the significance of 'technology in diplomacy'. In that context I have emphasized the importance of the Internet and social and digital media — often referred to as *cyber, e-* or *i-* diplomacy — and explored the impact of *virtuality* on the structure and operations of foreign ministries (Copeland, 2009). In this instance, however, I will shift the focus to an examination of what might be referred to as 'diplomacy *in* technology'. More precisely, I would like to examine a subject which dominated the news throughout much of 2010/11 — the WikiLeaks 'Cablegate' episode — with a focus on the implications for diplomacy and international relations.

The publication, between February 2010 and September 2011, of hundreds of thousands of US-origin diplomatic cables on the WikiLeaks web site revealed little about science, but much about diplomacy and technology. The cables were produced between December 1966 and February 2010 by 274 American diplomatic missions worldwide.[6] The classification of the messages varies from unclassified to secret, and they cover a vast array of subjects. It is the largest unauthorized transfer of government-origin classified information ever recorded.

As is so often the case in evaluating the impacts of technology, the implications of the WikiLeaks 'Cablegate' disclosures cut all ways. That is, the 'Cablegate' messages illustrate again the striking role of science and technology (S&T) as a two-edged sword in age of globalization. The same digital information and communications technologies which allowed the State Department — and the Pentagon, intelligence agencies, and other departments and agencies of the US government — to transmit, aggregate

---

[6] See WikiLeaks, http://www.wikileaks.ch/cablegate.html. *The Guardian* has been highly comprehensive and proficient in covering all aspects of the WikiLeaks story. See http://www.guardian.co.uk/media/wikileaks. Book length treatments are offered, among others, by Sifry and Rasiej (2011); Leigh and Harding (2011); and Mitchell (2011). For a comprehensive survey of initial US media coverage of 'Cablegate', see McDougall (2011).

and archive the cables, also allowed PFC Bradley Manning,[7] now convicted of being the inside source of the material, and WikiLeaks founder Julian Assange to engage in its unauthorized duplication, storage and dissemination. As with its roles in defence, energy, agriculture, and the environment, S&T can produce the innovations which solve problems even as they generate an entire range of new global challenges.

In evaluating the consequences of these disclosures for diplomacy and international relations, it is less a case of judging their impact as 'good' or 'bad' than it is a matter of coming to terms with the notion that there are both negative and positive elements. It accordingly seems appropriate to construct something of a ledger.

First, however, a few ambient observations about this most extraordinary incident.

The Big Picture: Issues without Answers

Absorptive capacity and mediation

In the wake of the much heralded information revolution, to avoid overload most people need certain types of information to be mediated — organized, digested, triaged, summarized — before it can effectively be absorbed. A single batch of 257,000 unedited diplomatic cables and related messages is beyond the capacity of most individuals to review and process. This is probably why Assange turned to five of the largest media organizations in the world — *The New York Times, the Guardian, El Pais, Der Spiegel* and *Le Monde* to handle job. Moreover, recall that the 'Cablegate' message onslaught came on top of the previous release of 392,000 military reports on Iraq and 92,000 on Afghanistan. For all but large teams of dedicated full-time researchers, this is simply too much information to effectively process.

Accountability, responsibility and systemic breakdown

Mr Manning copied the classified 'Cablegate' material in several tranches onto a Lady Gaga DVD. He then transmitted that digitalized data to Julian Assange, who by cut publication deals with the media organizations named

[7] For a defence of Manning's actions, see Madar (2011).

above and in so doing leveraged his possession of the material to propel himself to celebrity. Those events have had consequences (Trevor, 2011), and they may well illustrate the element of immediate responsibility for the disclosures. Yet it must be asked: where is the more extended institutional and personal accountability in all of this? Who designed and approved the data management system that allowed low level operatives like Bradley Manning access to such sensitive information (Orr, 2013)? Mr Manning, moreover, was reportedly considered mentally unstable by his employer, had been demoted for punching a woman officer in the face, and was about to be discharged. Given that there were apparently hundreds of thousands of people with similar clearance and access, Mr Manning could be considered a scapegoat for an event which was virtually inevitable. In that case, his conviction amounts to shooting the messenger.

The *9/11 Commission Report* (US Government, 2004) identified a lack of co-ordination among law enforcement and security agencies, and criticized the inadequate sharing of intelligence between organizations that resembled sealed information silos. Clearly, remedial action was required, but moving from an overly restrictive interpretation of the need to know to a free-for-all with no control over the copying and distribution of secret information amounts to an accident waiting to happen. This inexcusable failure of oversight represents a colossal lack of judgement and discretion at senior management levels.

Media play and profile

When the 'Cablegate' story first broke in February 2010, it elicited a round of over the top reactions from some American politicians and pundits — ranging from calls for Mr Assange's assassination (Fox News pundits) to demands that he be 'hunted down' (Sarah Palin). Coverage of the message *content* continued for a year and a half, but in the early stages of the story that coverage fell off radically. In the face of the relentless pressures generated by the immensely competitive and fragmented 24/7 news cycle, the centre of gravity of the 'Cablegate' affair seemed to shift. The story morphed from the substance of the disclosures to the tale of Saint Julian, the fearless crusader for press freedom. Next came a fascination with lurid reports of Assange's alleged sexploits in Stockholm, and then a prolonged focus on his heroic resistance to deportation from the UK. Assange's lawyers even raised the spectre of imprisonment at Guantanamo Bay, trial by military tribunal for terrorism, and possible execution (Addley, 2011).

This substitution of sensational human interest elements for hard news almost hijacked the narrative, and provides a dynamic illustration of the crisis of contemporary journalism — serious analysis is easily displaced by an emphasis on *infotainment*, and an ever greater preoccupation with *personalismo*.

The nature and function of WikiLeaks

Although more of a conduit than a source, WikiLeaks is nonetheless almost universally referred to as 'the whistle-blowing web site'. That description would not seem to apply in the case of the 'Cablegate' affair. Unlike WikiLeaks' presentation of the raw US military reporting contained in the Iraq or Afghanistan 'war logs',[8] or the Apache helicopter gun sight video footage of the 'death from above' killing of several civilians and two Reuters correspondents in Baghdad,[9] the 'Cablegate' data dump cannot be accurately described either as a 'leak', or as 'whistle blowing'. A leak is typically a single story, or at least a unified collection of documents; 'Cablegate' is an undifferentiated deluge. Whistle blowing typically refers to an act by which an employee, or a citizen, exposes corruption, illegality or wrong-doing within their organization or polity by those in authority. Although a limited number of 'Cablegate' messages report on apparent malfeasance abroad, it would be a stretch to maintain that the publication of that material constitutes whistle blowing. Candid diplomatic commentary and rigorous political analysis can of course be embarrassing to all parties if exposed, but that is something quite different.

Secondly, unlike the wildly popular Web 2.0 social media platforms such as Facebook, Twitter, YouTube and the ever expanding reaches of the

---

[8] See http://www.nytimes.com/interactive/world/war-logs.html for *The New York Times's* 'The War Log: An archive of classified military documents offers views of the wars in Iraq and Afghanistan' column.

[9] See Collateral Murder official website: http://www.collateralmurder.com/.

blogosphere, WikiLeaks.org and its spawn[10] more closely resemble old style, Web 1.0 'read only' and broadcast sites. To be sure, the material available is sometimes sensational or scurrilous, and by the spring of 2011 that kind of material was apparently being bundled and distributed strategically for maximum effect.[11] But neither the WikiLeaks site, nor its many mirror-image substitute sites are interactive — visitors cannot post or edit any of the material displayed. These sites, therefore, are not wikis.

If WikiLeaks/Cablegate is neither a wiki nor a leak, might it qualify as journalism? Most likely not. By virtue of its lack of breaking news, analytical or editorial content, the activities associated with the WikiLeaks site and its founder fall well outside of what is typically considered to constitute journalism. It may therefore seem curious that on 02 June 2011, Mr Assange was awarded the UK's coveted 2011 Martha Gelhorn Prize for journalism (Gunter, 2011). The actions of Mr Assange in receiving and disseminating the cables, and his defence of those actions in the world press, do not in my view compare favourably, for example, to the painstaking presentation, publication, and *analysis* of the 1600 'Palestine Papers' by *Al Jazeera* in January 2011.[12] That act falls within the purview of a responsible news organization, and differs markedly from entrepreneurial show-boating in support of technology-empowered celebrity.

In summary, Mr Assange may have pioneered a new form of highly individuated digital activism and political agency, but it would be inaccurate to describe his 'Cablegate' machinations as journalism, or whistle blowing, or leaking. That said, they have been costly — under pressure from the US government, PayPal, Amazon, MasterCard and Visa have stopped processing payments to WikiLeaks, while a series of Distributed Denial of Service attacks have taken the site off-line for extended periods.

---

[10] To circumvent the incapacitation of the main WikiLeaks site — during 2011/12 it was periodically off-line — a number of mirror sites have been established on different servers in order to ensure continuous access.

[11] On the colourful, controversial characterizations of leading Canadian politicians in advance of national elections in May 2011, see, for instance, Hildebrandt (2011).

[12] See http://english.aljazeera.net/palestinepapers/ for *Al Jazeera*'s 'The Palestine Papers' column.

And what of the larger consequences? They constitute somewhat of a mixed bag.[13]

Negative Impacts: Short-term Setbacks for Serving Envoys

A hit on tradecraft

By undermining the confidence, trust and respect upon diplomatic exchange is based, the revelations have introduced somewhat of a chill into diplomatic practice. Privacy has been invaded, and confidentially betrayed, both on a grand scale. As a result, new sources may hesitate to come forward. There have been reports of diplomats being excluded from high level political meetings for fear that private conversations may end up on the front pages.[14] The result is damage to networks, contacts, relationships. Keeping envoys isolated in their offices, out in the corridor, or on the other side of closed doors will necessarily affect the quality of reporting, analysis and, ultimately, decision-making.

Redoubled secrecy

Concerns over confidentiality will almost certainly lead to higher levels of classification, to less information sharing, and to a return to return to bureaucratic stovepipes and silos. Sensitive conversations are likely to go 'off paper' to secure telephony and face to face encounters. Fewer records of such exchanges will be made or retained, and that can only have a deleterious effect on governance.[15]

Counting the casualties

The disclosures have caused some collateral damage, including the expulsion of the US Ambassador to Ecuador, the resignation of the US Ambassador to Mexico, and — ironically — the firing of the Director General of Al Jazeera

---

[13]The following discussion was first presented at Otago University's Science Diplomacy Conference in June 2011, and was further developed as a chapter entitled *Digital Technology* (Copeland, 2013).

[14]Verified by the author through personal and confidential communications with serving diplomats.

[15]On the irony of WikiLeaks disclosures leading to greater secrecy in government, see Keller (2012).

News.[16] More worrisome still, despite the considerable effort to remove names and other possible identifiers on the part the large media organizations whom initially partnered with WikiLeaks and Mr Assange in the release of the cables, the entire unredacted trove was released on 1 September 2011, leading to acrimonious exchanges over who was to blame.[17] As a result, some sensitive, vulnerable sources were exposed (Shane, 2011). While the downside of this irresponsible action does not to date appear to have been dire as initially feared, it could easily have been tragic.[18]

In short, while worst-case scenarios have not been realized, the craft of diplomacy, the quality of public administration and governance, and several careers have suffered. The business of government, however, goes on, and the need to transact that business through international political communications endures. When it comes to the conduct of relations between states, there is often no alternative to direct contact. The means will evolve, and work-arounds will be found, but in the end, the diplomatic process will continue.

Positive Impacts: Many and Unexpected

For reasons suggested above, most assessments of the larger implications for diplomacy of the 'Cablegate' imbroglio have been negative. I believe, however, that the upside of this episode — a striking illustration of unintended consequences — has in fact been more significant.

Reinforcing honesty, consistency and transparency in government

Disclosures of this sort are becoming more frequent. As a result of the increased civic awareness and media oversight which has been engendered

[16] Wadah Khanfar was allegedly fired because of his susceptibility to U.S. editorial influence as described in a number of cables released by WikiLeaks in September 2011. For an assessment of these US lobbying efforts, see Forte (2011).

[17] The relationship between WikiLeaks and *The Guardian* has been ruptured over this issue. See, for instance, Cheng (2011).

[18] The US government reportedly had to relocate several sources, and one Ethiopian journalist, whose name appeared as a confidential contact in one of the cables, reportedly fled Addis Ababa, fearing his personal safety. See Committee to Protect Journalists (CPJ, 2011).

by such releases, governments are likely to be more careful to ensure that public statements align with facts gathered and actions taken. However much resented, as more whistles are blown — witness the more recent Edward Snowden affair, about which more later — the performance of politicians, departments and officials is likely to improve.

Increasing international affairs content in journalism

At a time of diminishing foreign and international coverage in the mainstream media,[19] the simple existence of this type of story has had a tonic effect on the quality of the news mix. Stories involving otherwise exotic locales such as Yemen, Libya, Tunisia, Egypt and Pakistan have hit the front pages. By injecting a large dose of international content, and bolstering its prominence, the usual preoccupation with local news, domestic issues and personalities has been leavened. In the age of infotainment, this re-balancing — however fleeting — can only be beneficial to the health of the body politic.

A new resource for scholars

For students of diplomacy and internal relations, publication of the quarter million plus 'Cablegate' messages has added substantially to an ever-growing e-collection of previously protected government documents. Elaborate screening mechanisms and protracted wait times — typically 25–50 years for documents of this classification — have been circumvented. This new archive represents a bonanza for journalists and scholars, one which offers telling, and highly contemporary insights into the nature of power and the exercise of influence. The advent of universal, free access to a research trove of this exceptional nature is unprecedented.

More attention to privacy and information protection

The magnitude of this breach may result in some technical and procedural improvements to communications security and innovations in the handling,

---

[19]This is a result of cost-cutting, media fragmentation and reader migration to new sources of information the Web. See, for example, Enda (2011).

storage and distribution of classified material.[20] In the case of genuinely sensitive content or sources, reforms of this variety can likely be justified, especially if not overzealously or inappropriately applied.[21]

Rehabilitation of the diplomatic brand

However ironic, the most notable effect of 'Cablegate' has been to bolster the diplomatic brand. Diplomacy suffers from a debilitating negative image. The mainstream view of diplomats and diplomacy is probably not far from a composite characterization which I have constructed based upon several years of informal focus group testing with cabbies in London, England. Their view? Diplomats are dithering dandies, lost hopelessly in a haze of irrelevance, stumbling blindly between protocol and alcohol...

The publication of hundreds of thousands of cables has subverted that corrosive caricature, and in so doing has burnished diplomacy's badly tarnished reputation. By illuminating the day-to-day reality of a very busy profession, the popular perception of envoys snoozing away their afternoons after long, well-lubricated lunches, or breezing around as privileged passengers in embassy limos, or drifting aimlessly through elegant receptions and lavish dinner parties has been punctured.[22] To the contrary, the 'Cablegate' dispatches show diplomats, time and again, working hard at their jobs, pursuing interests, projecting values and advocating policies. Many will find those values, policies and interests disagreeable, and in some cases extremely so. But the overwhelming picture which emerges is that of dedicated employees working long hours with their noses to the grindstone.

In the USA, this counter-cultural characterization has changed the minds of more than a few opinion leaders about role played and value added

---

[20] After publication of the *9/11 Report*, through a facility called SIPRnet, the State Department made its confidential reporting accessible to some 500,000 US government employees worldwide. This represented a severe over-reaction, and has since been scaled back.

[21] Well before Edward Snowden's revelations, cyber security had already become a growth industry, in part because of the increasing incidence of cyber spying. See, for example, Charkow (2011).

[22] There is legitimate representational work to be done in social settings. In most places, however, not least due to cost pressures, this sort of activity occupies an ever smaller proportion of a long work day.

by the State Department.[23] That alone could pay dividends at a time of increasing competition for scarce resources. Moreover, the sheer volume of reportage on WikiLeaks/Cablegate has had the effect of helping to bring diplomats and diplomacy from the farthest reaches of popular consciousness into something approximating the cultural mainstream. This de-mystification can have beneficial implications as regards diplomacy's brand *vis-a-vis* its international policy rivals.

Changing the Game: A Napster Moment for Government

What to make of it all?

Although the implications have been far-reaching, it is by no means clear that the 'Cablegate' disclosures were intended to support freedom of information, transparency, probity in government or defence of the public interest. Instead of serving as a conduit for the transmission of vital knowledge out of the shadows and into the light, this affair seems to have been more about personal self-aggrandisement and the commoditization of information. In the US, UK, Canada, Peru, Australia, India, Holland and elsewhere, releases seem to have been carefully timed and targeted, designed to produce maximum publicity for the source. This is closer to classic muck-raking and entrepreneurship than journalism, heroism or principled support for good governance. Some scepticism is clearly warranted regarding Mr Assange's claims that the 'Cablegate' revelations played a major role in encouraging of the Arab Spring.[24] While there was quite possibly some influence on the margins, it is also likely that few of those who participated in the uprisings had any clear idea of the content of the cables which reported on corruption, nepotism, and various other unsavoury practices in Tunisia and Egypt.

'Cablegate' may not have changed the world, but it has nonetheless produced a 'Napster moment' for governments, and disclosures of this sort may yet prove pivotal for international relations writ large. Just as the emergence

[23] Private conversations over the past year with politicians, academics and journalists have revealed a pattern of consistent admiration for the quality of the 'Cablegate' reporting, and several respondents admitted that their previously negative views of diplomacy had changed.

[24] Mr Assange's parody of a MasterCard commercial, designed to draw attention to the banking blockade against donations to WikiLeaks, is nonetheless brilliant. See http://www.youtube.com/watch?v=jzMN2c24Y1s.

of the music file sharing site Napster in the mid-1990s transformed the music retailing industry forever, the emergence of WikiLeaks, and the similar sites that are popping up all over cyberspace, looks very much like a game changer.[25] Think culture shift, with the Web emerging as a new political centre. The classified information monopoly once enjoyed by governments is over, and for those inclined towards secrecy and information control, life will never be the same again.[26]

Not least, the 'Cablegate' collection offers some compelling insights into machinations of US foreign policy. While it is impossible to know what percentage of total US diplomatic communications is represented by this sample, in the most contemporary messages, dated 2008–10, clear reporting priorities, such as the global financial crisis, climate change or the implications of power shifting to the Asia Pacific, are not much in evidence. S&T reporting in general, and reporting on science diplomacy initiatives in particular is under-represented given the looming nature of the threats and challenges.[27]

Viewed in aggregate, these reports suggest the antithesis of American grand strategy. Rather than providing a portrait of an empire at top of its game, the impression is one of a rather dishevelled Uncle Sam bumping along into the imperial darkness, desperately trying to plug cracks in an increasing number of failing dykes, world-wide. For conspiracy theorists, who see the dark side of American power behind everything that goes wrong in the world, this record offers little solace. Indeed, the content of the 'Cablegate' archive will more likely be interpreted by the declinist school as indicative of America's ebbing place in the world.

The Wrap: Fewer Bullets, More Bytes

Information generated by civil servants has been financed by taxpayers and should therefore be available in the public domain. Transparency in government

[25] It has certainly given a boost to 'hactivism', and drawn attention to the activities of shadowy groups such as Anonymous. See Milan (2011).

[26] This lesson may take some time to sink in. The US government, for example, has ordered its employees to refrain from accessing web sites hosting the WikiLeaks cables. This has produced some Kafkaesque situations. See, for example, Van Buren (2011).

[27] Of the total of over 250,000 cables, only 8,627 contained any reference to science or technology.

is important. But not all information needs to be freed. To give just one example, publication early on in the 'Cablegate' affair of the US government's estimate of the world's most vulnerable critical infrastructure sites surely did not serve the general interest (Spector, 2010). Even if the details on such locations were otherwise available, an estimate of the American government's foremost concerns was not. In any event, why make high grade research material available to those who might use it to do harm?

In a similar vein, why risk the exposure of sensitive contacts? And how many future sources of valuable intelligence will not now come forward for fear of being revealed and punished?

The 'Cablegate' episode has happened not so much because it *should* — to repeat, there is little evidence of probity or the public interest in play as motives — but because it *could*. The issues and key drivers seem to have more to do with personal ambition, digital capacity and technological possibility than with morals, ethics or the people's right to know.

To conclude: while I am not entirely convinced that the world needs more 'Cablegates', I am certain that contemporary international relations, and in particular the prospects for development and security, would benefit from more science diplomacy. There is a daunting performance gap at present, and in an increasingly *heteropolar*[28] world no amount of armed force can fill it. The dispatch of an expeditionary force will not permit citizens to occupy the alternatives to the carbon economy. The most lethally equipped military cannot defend national borders against pandemic disease. Air strikes are ineffective in the battle to reverse climate change.

Today, the lion's share of international policy resources are vested in defence, but the more profound threats are elsewhere. That disconnect constitutes an elemental obstacle to progress. Innovative thinking, more effective communications and a fundamental shift in direction are required.

The very act of international scientific and technological cooperation encourages policy harmonization and action in concert. Moreover, the

[28] Defined as: an emerging world system in which competing states or groups of states derive their relative power and influence from dissimilar sources — social, economic, political, military, cultural. The disparate vectors which empower these heterogeneous poles are difficult to compare or measure; stability in the age of globalization will therefore depend largely upon the diplomatic functions of knowledge-driven problem solving and complex balancing. See Copeland (2012).

intersection of science, technology and international policy represents a strategic nexus where this transformational brand of public diplomacy — one which is powered by continuous learning in order to solve problems constructively and without recourse to violence — could make all the difference. As much as anything else, however, the dearth of science diplomacy reporting in the WikiLeaks corpus underscores SD's relative scarcity.

'Cablegate' aside, there are some bright spots on the horizon, such as the OECD initiative on Science, Technology, Innovation and Global Challenges (Koch, 2011; OECD, 2012) and the most recent G8 Science Ministers Statement (UK Government, 2013). If the public can be engaged, and the leadership and political will summoned to provide the requisite financial and human resources, I am convinced that an investment in science diplomacy would pay handsome returns.

'Cablegate' has demonstrated that diplomacy can deliver. Given the parlous state of our world, racing as it is towards some still undefined tipping point beyond which recovery may be difficult if not impossible, that is message worth sending.

Epilogue: Ambiguous Outcomes and a Whirled View

The core of the assessment presented above was prepared for an international conference on science diplomacy organized by the University of Otago, New Zealand. in June 2011. Since that time, Julian Assange has been granted protection by the Ecuadorean government, but as a fugitive he faces arrest and deportation if he leaves the sanctuary of their Embassy in London.[29]

In the wake of 'Cablegate', it seems deeply ironic that Mr Assange has found it necessary to cast his fate upon the tender mercies diplomatic convention.

Bradley — now Chelsea — Manning was acquitted on the most serious charges of 'aiding the enemy', but was found guilty of espionage, computer fraud, possession of restricted documents, and theft, and sentenced to 35 years in prison, a record for this sort of offense.[30]

---

[29] For Assange's take on WikiLeaks and Cablegate, see Assange (2012).

[30] The detailed *Wikipedia* entry on Bradley/Chelsea Manning is very useful. See http://en.wikipedia.org/wiki/Bradley_Manning.

In a statement about why he revealed the documents, Mr Manning said he acted to expose American diplomatic deceit ('Cablegate'), as well as the U.S military's 'bloodlust' and disregard for human life (Iraq and Afghanistan war logs). He claimed that he wanted to start a debate on foreign and defence policy, and that he chose information that was dated and would not the harm the interests of the United States (Combs, 2013). The US government maintains that Mr Manning's disclosures flouted the law and have seriously compromised security.[31]

For his part, President Obama continues to surprise those who were expecting more openness and reform in his second term. When Mr Manning pleaded guilty earlier this year to several lesser offences that would have brought him about 20 years of imprisonment, the government refused to bargain and opted instead to prosecute the most serious charges. In fact, since taking office less than six years ago, Obama has pursued more espionage charges against government employees than all other past presidents combined (Mattingly and Nichols, 2012).

Some two years after 'Cablegate', NSA contractor turned cyber surveillance whistleblower Edward Snowden has been granted temporary asylum in Russia. His disclosure of documents detailing mass telephone and internet monitoring, commercial code-breaking, and a vast array technological eavesdropping by the intelligence agencies of the United States, Britain, Australia and France, often with active private-sector collusion, may represent the most significant leaks in US history.[32] Subsequent reporting has provided crucial information regarding the actions of the secret Foreign Intelligence Surveillance Court, whose classified rulings vested the National Security Agency with sweeping new powers (Lichtblau, 2013).

Like Manning, Snowden has been proclaimed a hero in some quarters and a traitor in others (Cassidy, 2013). The US government has charged him with the theft of government property, unauthorized communication of national defence information, and willful communication of classified intelligence to unauthorized persons. They are seeking his extradition, and have criticized the governments of Russia and China for their failure to co-operate.

[31] For a critique of the government's handling of the case, see Hedge (2013).

[32] See the anthology presented in 'The NSA Files' by *The Guardian*. Available at: http://www.theguardian.com/world/the-nsa-files.

Washington appears intent on sparing no effort in its attempt to ensure that Mr Snowden's fate resembles something closer to that of Bradley Manning than that of Julian Assange.

By throwing the book at, and making an example of Bradley Manning, Edward Snowden and all other actual and potential whistle-blowers, the US administration clearly hopes to induce an intimidating chill throughout the civil service, and in so doing reduce the incidence of leaking.

Such efforts are unlikely to succeed (Engelhardt, 2013).

More disturbing is the question of whether or not this strategy (Pitzke, 2013), in combination with rising inequality[33] and the imposition of curbs on civil liberties and constitutional rights (ACLU, n.d.), is transforming the erstwhile land of the free into a something disturbingly Orwellian — a national security state.

In that respect, Snowden's continuing string of revelations may encourage debate on the growing tension between divergent public policy objectives. By illuminating the extent of state-sponsored cyber-spying directed at both domestic and foreign targets, Mr Snowden's disclosures have set the stage for possible remedial action, in the US Congress or elsewhere.

While it will not be easy, the search for a more judicious balance between the competing imperatives of information privacy and individual rights on the one hand, versus national security and counterterrorism on the other, is long overdue.

Further afield, the targeting of foreign governments (Brooks, 2013) and international organizations (Deighton, 2013) by the NSA may be unsurprising, but having the details appear on front pages worldwide makes for rough diplomatic sailing. Equally damaging: Under the rubric of *twenty-first century statecraft* (US Department of State, n.d.) one of the principal objectives of US international policy in recent years has been the promotion of internet freedom, access and openness (Dickinson, 2010). The credibility of American advocacy related to those themes, already strained (Schulman, 2011), and has now been shredded.

All told, this debilitating widening of 'say-do gap' (Copeland, 2009) can only make matters worse in the world for the struggling superpower.

Final thoughts?

---

[33] Recall the Occupy Wall Street movement, and see Graham (2013).

Julian Assange, Bradley Manning and Edward Snowden have attained global celebrity through the innovative use of digital technology. In so doing they have influenced international relations in a fashion which would have been unimaginable only a few years ago (Logan, 2013).

But…Assange is in a gilded cage, locked down and angry. Manning is facing long years in prison. Snowden, following Assange, may well land lucrative book and film deals (Toor, 2010), yet the prospect of life on the run, or of permanent exile, cannot be very appealing.

While the public interest may ultimately emerge as a net beneficiary, the personal costs attached to the pursuit of this formula for individual self-empowerment have been extreme.

References

American Civil Liberties Union (ACLU) (n.d.) *Reform the Patriot Act*, [Online], Available: https://www.aclu.org/reform-patriot-act.

Addley, E. (2011) 'WikiLeaks: Julian Assange faces execution or Guantonamo detention', *The Guardian*, 11 January, [Online], Available: http://www.theguardian.com/media/2011/jan/11/julian-assange-wikileaks-execution-gantanamo.

Assange, J. (2012) 'Two years of Cablegate as Bradley Manning testifies for the first time', *Huffington Post*, 29 November, [Online], Available: http://www.huffingtonpost.com/julian-assange/wikileaks-bradley-manning-testifies-cablegate_b_2215387.html.

Brooks, B. (2013) 'Brazil opens investigation into US spying', *NewEurope Online*, 8 July, [Online], Available: http://www.neurope.eu/news/wire/brazil-opens-investigation-reports-us-spying-whether-telecoms-brazil-aided.

Cassidy, J. (2013) 'Why Edward Snowden is a hero', *The New Yorker*, 10 June, [Online], Available: http://www.newyorker.com/online/blogs/johncassidy/2013/06/why-edward-snowden-is-a-hero.html.

Charkow, R. (2011) 'Cyber spying is the new face of espionage', *CBC News*, 21 September, [Online], Available: http://www.cbc.ca/news/canada/story/2011/09/20/f-cyber-espionage.html.

Cheng, J. (2011) 'WikiLeaks: Unredacted cable release is Guardian's fault', *Ars Technica*, 2 September 2011, [Online], Available: http://arstechnica.com/tech-policy/news/2011/09/wikileaks-unredacted-cable-release-is-guardians-fault.ars.

Combs, D. (2013) 'PFC Manning's Written Statement in Support of his Guilty Plea', *Law Office of David E. Coombs*, 4 March, [Online], Available: http://www.army-courtmartialdefense.info/2013/03/pfc-mannings-written-statement-in.html.

Copeland, D. (2009) 'PD's Most Formidable Adversary', *The CPD Blog*, 16 June, [Online], Available: http://uscpublicdiplomacy.org/index.php/newswire/cpdblog_detail/pds_most_formidable_adversary_the_say_do_gap/.

Copeland, D. (2009) 'Virtuality, Diplomacy and the Foreign Ministry: Does foreign affairs and international trade Canada need a "V Tower"', *Canadian Foreign Policy*, 15(2): 1–15.

Copeland, D. (2012) 'Heteroplarity, globalization and the new threat set', *EmbassyMag*, 21 February, [Online], Available: http://www.embassymag.ca/dailyupdate/view/199.

Copeland, D. (2013) 'Digital Technology', in Cooper, A., Heine, J. and Thakur, R. (ed.) *The Oxford Handbook of Modern Diplomacy*, Oxford: Oxford University Press.

Committee to Protect Journalists (CPJ) (2011) 'Ethiopian journalist ID'd in WikiLeaks cable flees country', *CPJ*, 11 September, [Online], Available: http://cpj.org/x/45d1.

Deighton, B. (2013) 'EU confronts US over alledged spying on European allies', *Globe and Mail*, 30 June, [Online], Available: http://www.theglobeandmail.com/news/national/eu-confronts-us-over-alleged-spying-on-european-allies/article12899295/.

Dickinson, E. (2010) 'Internet Freedom', *Foreign Policy*, 21 January, [Online], Available: http://www.foreignpolicy.com/articles/2010/01/21/internet_freedom.

Enda, J. (2011) 'Retreating from the World', *AJR Archives*, December/January, [Online], Available: http://www.ajr.org/article.asp?id=4985.

Engelhardt, T. (2013) 'Letter to an unknowed whistleblower', *TomDispatch*, 17 September, [Online], Available: http://www.tomdispatch.com/post/175748/tomgram%3A_engelhardt%2C_how_to_build_a_national_security_blowback_machine/#more.

Forte, M. (2011) 'What WikiLeaks' US Embassy Cables Reveal about US Pressure and Propaganda', *ZNet*, 25 September, [Online], Available: http://www.zcommunications.org/what-wikileaks-u-s-embassy-cables-reveal-about-u-s-pressure-and-propaganda-by-maximilian-forte.

Graham, C. (2013) 'America's broken dream', *Project Syndicate*, 5 September, [Online], Available: http://www.project-syndicate.org/commentary/the-global-impact-of-rising-inequality-in-the-us-by-carol-graham.

Gunter, J. (2011) 'Julian Assange wins Martha Gellhom Prize for Journalism', Journalism.co.uk, 2 June, [Online], Available: http://www.journalism.co.uk/news/julian-assange-wins-martha-gellhorn-prize-for-journalism/s2/a544492/.

Hedge, C. (2013) 'Bradley Manning and the Gangster State', *TruthDig*, 21 August, [Online], Available: http://www.truthdig.com/report/item/bradley_manning_and_the_gangster_state_20130821/#.UhvJ41d0em0.email.

Hildebrandt, A. (2011) 'US cables dissect Canadian leaders: WikiLeaks', CBC News, 1 May, [Online], Available: http://www.cbc.ca/news/world/story/2011/05/01/wikileaks-canada-leaders.html.

Keller, B. (2012) *WikiLeaks, a Postscript*, 19 February, [Online], Available: HYPERLINK "http://www.nytimes.com/2012/02/20/opinion/keller-wikileaks-a-postscript.html" http://www.nytimes.com/2012/02/20/opinion/keller-wikileaks-a-postscript. html.

Koch, P. (2011) *Presentation to the OECD Steering Group for Governance of International Cooperation on Science, Technology and Innovation for Global Challenges*, 13 October, [Online], Available: http://www.slideshare.net/perkoch/oecd-stig-october-2011-chairs-presentation.

Leigh, D. and Harding, L. (2011) *WikiLeaks: Inside Julian Assange's War on Secrecy*, New York: Public Affairs.

Lichtblau, E. (2013) 'In secret, court vastly broadens powers of NSA', *New York Times*, 6 July, [Online], Available: http://www.nytimes.com/2013/07/07/us/in-secret-court-vastly-broadens-powers-of-nsa.html?_r=3&.

Logan, S. (2013) 'Has Snowden left international relations stuck in a transit lounge', *EastAsiaForum*, 11 July, [Online], Available: http://www.eastasiaforum.org/2013/07/11/has-snowden-left-international-relations-stuck-in-a-transit-lounge/.

Lord, K.M. and Turekin, V.C. (2007) 'Time for a New Era of Science Diplomacy', *Science*, 315(5813): 769–770.

Madar, C. (2011) 'Why Bradley Manning Is a Patriot, Not a Criminal', *TomDispatch*, 10 February, [Online], Available: http://www.tomdispatch.com/archive/175352/chase_madar_the_trials_of_bradley_manning.

Mattingly, P. and Nichols, H. (2012) *Obama purseing leakers sends warning to whistle-blowers*, 17 October, [Online], Available: http://www.bloomberg.com/news/2012-10-18/obama-pursuing-leakers-sends-warning-to-whistle-blowers.html.

McDougall, D. (2011) 'Wikileaks Cablegate Media Monitor Report', *The CPD Blog*, 17 May, [Online], Available: http://uscpublicdiplomacy.org/pdin_monitor_article/wikileaks_cablegate_media_monitor_report.

Milan, S. (2011) 'One year after Cablegate: WikiLeaks legacy on cyber-activism', *CitizenLab*, 29 November, [Online], Available: https://citizenlab.org/2011/11/one-year-after-cablegate-wikileaks%E2%80%99-legacy-on-cyberactivism/.

Mitchell, G. (2011) *The Age of WikiLeaks: From Collateral Murder to Cablegate (and Beyond)*, Dayton, OH: Sinclair Books.

Nye, J. (2004) *Soft Power: The Means to Success in World Politics*, New York: Public Affairs.

Organization for Economic Cooperation and Development (OECD) (2012) *Meeting Global Challenges through Better Governance: International Co-operation in Science, Technology and Innovation*, [Online], Available: http://www.oecd.org/

sti/sci-tech/meetingglobalchallengesthroughbettergovernanceinternationalco-operationinsciencetechnologyandinnovation.htm.

Orr, B. (2013) 'How did low-level employees access national secrets?', *CBS News*, 30 July, [Online], Available: http://www.cbsnews.com/8301-18563_162-57596218/how-did-low-level-employees-access-national-secrets/.

Pitzke, M. (2013) 'War on whistle-blowers', *Der Spiegel*, 24 July, [Online], Available: http://www.spiegel.de/international/world/obama-wages-war-on-whistleblowers-and-journalists-a-912852.html.

Schulman, M. (2011) 'The State Department's shameful record on Internet freedom', *NewRepublic*, 8 August, [Online], Available: http://www.newrepublic.com/article/politics/93283/state-department-internet-freedom-china-censorship?page=0,1.

Shane, S. (2011) 'WikiLeaks leaves names of diplomatic sources in cables', *New York Times,* 29 August, [Online], Available: http://www.nytimes.com/2011/08/30/us/30wikileaks.html?_r=1&hpw.

Sifry, M. and Rasiej, A. (2011) *WikiLeaks and the Age of Transparency*, Berkeley, CA: Counterpoint.

Smith, C.L. (2012) 'Synchrotron light and the Middle East: Bringing the region's scientific communities together through SESAME', *Science and Diplomacy*, 16 November, [Online], Available: http://www.sciencediplomacy.org/perspective/2012/synchrotron-light-and-middle-east.

Spector, N. (2010) 'WikiLeaks mad attack on Canada', *Globe and Mail*, 6 December, [Online], Available: http://www.theglobeandmail.com/news/politics/second-reading/spector-vision/wikileakss-mad-attack-on-canada/article1826060/.

The Royal Society (2010) *New Frontiers in Science Diplomacy: Navigating the Changing Balance of Power,* January, [Online], Available: https://royalsociety.org/~/media/Royal_Society_Content/policy/publications/2010/4294969468.pdf.

Toor, A. (2010) 'Julian Assange needs a $1.5 million book deal to pay his lawyers', *SWITCHED*, 27 December, [Online], Available: http://www.switched.com/2010/12/27/julian-assange-1-5-million-book-deal/.

Trevor, T. (2011) 'Cablegate One Year Later', *Electronic Frontier Foundation*, 28 November, [Online], Available: https://www.eff.org/deeplinks/2011/11/cablegate-one-year-later-how-wikileaks-has-influenced-foreign-policy-journalism.

TVNZ (2011) *Scientists top 'most trusted' list*, 20 June, [Online], Available: http://tvnz.co.nz/national-news/scientists-top-most-trusted-list-4247442.

UK Government (2013) *G8 Science Ministers Statement*, 12 June, [Online], Available: https://www.gov.uk/government/publications/g8-science-ministers-statement-london-12-june-2013.

US Department of State (n.d.) *21st Century Statecraft*, [Online], Available: http://www.state.gov/statecraft/overview/.

US Government (2004) *The 9/11 Commission Report*, 22 July, [Online], Available: http://govinfo.library.unt.edu/911/report/911Report.pdf.

Van Buren, P. (2011) 'Freedom Isn't Free at the State Department', *TomDispatch*, 27 September, [Online], Available: http://www.tomdispatch.com/post/175446/tomgram%3A_peter_van_buren%2C_wikileaked_at_the_state_department/#more.

Zaharna, R. (2010) *From Battles to Bridge: US Strategic Communications and Public Diplomacy After 9/11*, New York: Macmillan.

Part III

Science for Diplomacy: Using Science Co-operation to Improve Relations between Countries

CHAPTER 10

Triangulating Science, Security and Society: Science Cooperation and International Security

Jeffrey Boutwell

This chapter will focus on the various ways in which science cooperation among nations and peoples can help improve international relations and security. In the past, certainly through much of the 20th century, modes of scientific cooperation more often than not took the form of joint governmental science projects and/or individual contacts and research work among members of the international scientific community. Through membership in international scientific associations and participation in large-scale scientific projects that were beyond the means of any one government, such as the International Space Station or CERN, the high energy physics research facility near Geneva, scientists were truly members of an international community that often transcended national borders and identity.

Today, however, with personal communications and social media technologies having revolutionized the way in which people around the world interact and instantaneously spread information, the modes of scientific cooperation have expanded well beyond the 'formal' scientific community. Individual scientists find it far easier to communicate directly with colleagues in far-flung locales, thus facilitating the growth of more informal networks for the sharing of information.

More broadly, a much larger proportion of society is now empowered to join the debate over the ways in which science impacts both physical and human security. One well-known example is that of the controversial public debate over the role to be played by genetically modified organisms (GMOs) in shaping our food supply and food security, and how this debate plays out very differently in regions such as Europe, the United States, Latin America and Asia.

There's little question that there will be both benefits and drawbacks to society being empowered through this spread of communications technologies and social media to play a greater role in shaping science and security public policy choices. Looking again at the GMO/food security issue, greater public awareness of the role of genetically-modified crops is to be welcomed. Yet, when terms such as 'Frankenfoods' are promoted by opponents of GMOs, it becomes difficult to argue the scientific merits of the issue and evaluate the extent to which GMOs could be a legitimate tool in promoting food security, particularly in developing nations.

To evaluate how international cooperation and public involvement in science and security issues has evolved from the final decades of the 20th century to the new world we face in the 21st century, I will focus on three issues in particular: missile defence and nuclear deterrence; civilian and military uses of outer space; and global climate change in general and changes to the Arctic polar region in particular. The first two issues fall into the category of more traditional defence security issues among nations, though each did have an important public society component, while the third issue, climate change, more directly raises issues of human security and how we as societies adapt to greatly changed environmental and living conditions through the remainder of the 21st century and beyond.

Society as Interlocutor between Science and Security

As is evident in many of the chapters in this volume, especially those that emphasize the role of new communications media that disseminates science and technology information to a far wider audience than ever before, Science Diplomacy per se is no longer the sole preserve of the scientific and public policy communities. Whether the issue is global climate change, nuclear weapons and deterrence, cooperation and conflict in outer space, or environmental and food security, the educated publics of the world are far more knowledgeable about such issues than was the case 50 years ago during the Cold War.

The organization that I worked with for 30 years, the Pugwash Conferences on Science and World Affairs, has changed greatly since it was founded in the fishing village of Pugwash, Nova Scotia in July 1957 (Rotblat, 1972). At that first meeting, 22 eminent physicists from the US, Russia, Europe and Asia met to discuss the threat posed by nuclear weapons, particularly the then newly-tested thermonuclear, hydrogen bombs. The discussions they had over three days covered such topics as the radiation hazards posed by both the peaceful and military uses of atomic energy, the control of nuclear weapons, and the social responsibility of scientists. Particularly on the issue of atomic radiation, it is unlikely that, in the 1950s, more than a few hundred scientists around the world would have had the necessary expertise to understand, much less participate in, those discussions that ultimately gave birth to the Pugwash organization.

Today, global access to the very latest in scientific information and resources to those with a cellphone, computer, or iPad means that the discourse of international science and technology issues can take place much more widely both within and across societies.

Such ease of access does not, of course, automatically translate into a more rational discourse of such issues — as we've seen all too often in the contentious debates on global climate change[1] and genetically-modified food. To begin with, there will still be honest differences of opinion between scientists and policymakers on the interpretation of data and the significance of different models of the future. More than that, there will still be policymakers and opinion shapers who will either knowingly misuse scientific data, or dismiss scientific objectivity because of political, ideological or fundamentalist religious biases.

Nonetheless, important scientific and technological data, information, and resources are no longer the privileged preserve of just the scientists and policymakers. Informed citizens can now undertake their own analyses of important national and international security issues, and they can join together in public citizen movements to promote particular policies and points of view. In the US, there are a wide variety of organizations that seek to provide greater accessibility to scientific data and analyses and bring this information to bear on the making of public policy. Such groups, including the Union of Concerned Scientists, the Natural Resources Defence Council,

[1] For a good overview of the science and public discussion of the global climate change debate, see Pilkey *et al.* (2011).

and the Federation of American Scientists, to name just a few, are able to mobilize thousands if not tens of thousands of their members — both scientists and well-informed citizens — to speak out and become involved in important public policy issues. It is against this backdrop of greater informed public participation in Science Diplomacy, both actual and potential, that I want to cast my remarks on the three policy issues of missile defence and nuclear deterrence, the military uses of outer space, and Arctic security as affected by global climate change.

Before doing so, however, I will digress just a bit to emphasize the importance that individual scientists can have in influencing the policies of national governments on national and human security issues. Because of the international links that scientists have with their colleagues around the world, made all the easier by the revolution in communications technologies, there remains great scope so that even one individual can play a pivotal role in making society at large aware of threats to the human condition. This was certainly true in the past, as evidenced by the examples below, and will remain so in the future.

The Role of the Individual

Three scientists who exemplify the importance one individual can make are Joseph Rotblat of Poland, Matthew Meselson of the United States, and Hussain al-Shahristani of Iraq.

Having emigrated from Poland in the 1930s to work in the Liverpool physics lab of James Chadwick, Rotblat then joined the Manhattan Project in Los Alamos, New Mexico in 1943. Having lost his wife and other family members to Nazi death camps, Rotblat was convinced of the need to prevent Hitler from being the first to acquire atomic weapons. When it became clear by late 1944 that the Nazis were not going to be successful, however, Rotblat voluntarily resigned from the Los Alamos effort — for reasons of personal conscience — because of his deep misgivings over the catastrophic destructive power of atomic weapons. He returned to Britain to work on medical applications of nuclear technology, at great personal cost to his reputation and professional standing.

Jo Rotblat's essential credo as a physicist was that scientists must take responsibility for their work; it is not enough to abdicate that responsibility by invoking public policy or other political rationales. It was this principled stance that brought Rotblat to the attention of the British philosopher Bertrand Russell in the 1950s and helped launch the Russell-Einstein Manifesto of 1955 that alerted the international community to the dangers of thermo-nuclear weapons and in turn gave rise to the Pugwash Conferences on Science and World Affairs in Pugwash, Nova Scotia in 1957. Some 40 years later, in 1995, Rotblat and Pugwash shared the Nobel Peace Prize for their efforts to reduce and eliminate the threat posed by nuclear weapons.

Matthew Meselson is a professor of biochemistry at Harvard University, a longtime participant in Pugwash meetings, and one of the foremost experts in the world on chemical and biological weapons. In the 1980s he was centrally involved in the 'yellow rain' controversy, when the US government accused the Soviet Union of supplying T-2 *mycotoxins* to Vietnam and Laos for use in counter-insurgency warfare, allegedly resulting in the deaths of as many as 10,000 people. An independent team of scientific experts led by Meselson travelled to Southeast Asia and ultimately offered a much more compelling thesis for the 'yellow rain' droplets found on vegetation: that they were in fact honeybee faeces resulting from the defecation of digested pollen grains from large swarms of bees. Meselson and other scientists faced the enormous resources of the US government — which has never fully retracted its allegations, though it has acknowledged serious errors in its methodology and in the interviewing of local inhabitants — in seeking to provide factual scientific evidence about a very controversial issue at a time of great international tension during the Cold War (Tucker, 2001).

While Matthew Meselson had the resources and prestige of a Harvard professorship in his fight with the US government, the same was not true for Hussain al-Shahristani in his opposition to Saddam Hussein's desire to build an Iraqi nuclear weapons capability. As a nuclear physicist and chief scientific advisor for Iraq's civil nuclear program in the 1970s, Shahristani was imprisoned in 1979 for refusing to work on Saddam's nuclear weapons program. He spent 11 years in prison, much of it at the notorious Abu Ghraib prison, being tortured and in solitary confinement. He was promised his freedom by Saddam's half-brother and head of the secret police, Barzan al-Takriti, if he would recant and lead a covert effort to build nuclear weapons, but he refused. He remained in prison until 1990, escaping during the confusion during the first Gulf War. He made his way first to Iran, then to

England. In 2003, shortly after Saddam's fall during the second Gulf War, Shahristani became one of the first prominent Iraqi exiles to return to Baghdad, where he soon became Minister of Oil and an important figure in the political and economic reconstruction of the country (Salama and Hunter, 2005).

In thinking about these three individuals, it's instructive to remember the price each paid — personally and professionally — for opposing their governments and seeking to influence policy with their scientific expertise. In the case of Joseph Rotblat, he was placed under FBI surveillance and — in some quarters — accused of being a Soviet spy. Matthew Meselson became the bête noire of US Secretary of State Alexander Haig and was accused of undermining US security at a time of great tension between the US and Soviet Union. Hussain al-Sharistani, of course, paid the heaviest price — physical and mental torture — for refusing his government's command that he covertly developed nuclear weapons for Saddam Hussein.

Fortunately, the stories of all three of these scientists have had successful and positive outcomes, and the experiences of all three demonstrate the importance of the individual being responsible for his or her scientific work and utilizing it in the service of humanity.

Scientists and Ballistic Missile Defence

A good case study of how the scientific community can positively influence policy to reduce tensions and help restrain unrestrained arms competition is that of nuclear weapons and missile defences in the 1960s and 1970s. During that period, the Pugwash Conferences and other communities of scientists were involved in arranging influential meetings between U.S. and Soviet scientists on the controversial role of missile defence and whether it would contribute to or undermine nuclear deterrence and stability. At the beginning of these discussions, many Soviet scientists took what would seem to be the eminently reasonable stance that, in military affairs, 'offense is bad and defence is good'. They would ask, what can be morally wrong with missile defences (ABM, anti-ballistic missile, systems) whose only purpose is to shoot down incoming ballistic missiles armed with nuclear weapons that could incinerate cities and kill hundreds of thousands if not millions of people? There were also many in the US government and military making the same argument, and indeed the Johnson administration in the 1960s began deploying the Nike Zeus and Sentinel missile systems to protect American cities from Soviet nuclear-tipped missiles (Chayes and Weisner, 1969).[2]

What was missing from the calculus of the 1960s ABM debate was the reality that, because of the enormous destructive power of nuclear weapons, the US would need a success rate of well over 90 percent against incoming Soviet nuclear missiles to prevent catastrophic destruction to American cities and territory. Even if 10 out of 100 incoming Soviet warheads, carrying several hundred kiloton warheads, penetrated your defence, the result could be tens of millions dead and dying. In the language of the time, the task of the defensive missiles would be like 'a bullet hitting a bullet', intercepting missile re-entry vehicles traveling at seven kilometres per second and carrying sophisticated countermeasures with which to evade and confuse the ABM system. The final straw was the cost-benefit ratio, that building and deploying more offensive missiles was cheaper than building defensive missiles and radars, thus the Soviets could seek to overwhelm the US defence.[3]

As was argued in Pugwash and other such meetings at the time, the result of both the US and Soviets building defensive missile systems could well be an intensification of the offensive arms race, as each country would build more ICBMs with which to overwhelm the supposed effectiveness of the other's missile defences. American and Soviet scientists, having developed their analyses through the give-and-take of scientific meetings, argued that it would be more cost-effective, in terms of both security and military budgets, to place limits on offensive ICBMs and heavy bombers on the one hand, and missile defences on the other, so that neither side would feel the need to constantly keep expanding their nuclear arsenals. These were the arguments that ultimately persuaded Presidents Nixon and Brezhnev and which provided the foundation for the 1972 SALT I and ABM agreements that helped cap the nuclear arms race at the time.

To be sure, there will always be those seeking the magic of technological fixes to solve complex issues of public policy and national security, and the

---

[2] Chayes had been a legal advisor during the Kennedy administration, while Jerome Wiesner was Kennedy's chief Science Advisor and later President of the Massachusetts Institute of Technology.

[3] For a good overview of the technologies and inherent difficulties in achieving a successful missile defence system, see Bethe *et al.* (1986).

same has been true of missile defence. In the 1980s, President Reagan had his vision of Star Wars, spaced-based technologies (Yonas[4], 1986; Hafner, 1986), and more recently, the second President Bush began to deploy missile defences in Alaska and Europe to counter presumed threats from the Middle East.

There is little doubt that both American and Soviet scientists played a major role in the 1960s and 1970s in influencing their national policies on nuclear weapons and missile defences. Such scientists provided a credible counterweight to the military and defence industrial complex of both countries in demonstrating that building more weapons systems would not improve the security of either the United States or the Soviet Union. While public opinion did play a role (especially on the issue of nuclear atmospheric testing in the late 1950s and early 1960s), it was clearly secondary (Evangelista, 2002).

Today, I would argue, a major difference in debates over national security issues such as missile defence is that there exists a more informed public, able to mobilize through specialized organizations like the Union of Concerned Scientists (UCS) and Federation of American Scientists (FAS), that can provide alternative analyses on the pros and cons of spending billions of dollars on missile defences as a means of solving the threat posed by nuclear weapons. Granted, such scientific and civic groups have to contend with the 'iron triangle' of the military services, defence contractors and their supporters in Congress in debates over missile defence and other national security issues. As we saw during the Bush administration from 2000 to 2008, the defence lobby prevailed in allocating enormous resources to the research, development and initial deployment of new missile defense systems.

Nonetheless, it is true that the existence of well-organized scientific interest groups in the US helped raise the bar that missile defence proponents had to surmount in moving ahead with missile defence, even in the post 9/11 security-sensitive environment. Then, when President Obama took office in January 2009, many of the experts from the science and arms control communities who had been lobbying against a carte blanche for missile defences

---

[4] Yonas was the chief scientist of the Strategic Defence Initiative Organization at the Department of Defence during the Reagan administration.

became senior officials in the administration.[5] This made it easier for the White House to take on the Congressional proponents of missile defence and place important constraints on the program.

Weaponisation of Outer Space

A second major international security issue lying at the intersection of science, society and security is that of the potential weaponisation of outer space.

In 1967, the Outer Space Treaty laid the basic framework for establishing outer space (beyond the limits of the earth's atmosphere, known as the Karman Line and set somewhat arbitrarily at 100 kilometres above the earth), plus the Moon and other celestial bodies, as the "common heritage of mankind" which should be used exclusively for peaceful purposes and thus free of weaponisation and military competition. In the event, outer space was declared to be a nuclear-weapon free zone, yet many other military and conventional weapons activities were permitted, even if they have not been fully exploited to date (Stares, 1985).

In the years since, the US, Russia and China have conducted basic scientific research and testing of anti-satellite weapons, based on a variety of kinetic and laser technologies. Such systems have not yet been deployed, and attempts in the UN Conference on Disarmament in Geneva are still underway to begin negotiations on a treaty banning anti-satellite weapons, but with little success so far.

In a world that is increasingly dependent on space satellites for critically important communications, information, navigation, and geophysical and weather research, the prospect of 'an arms race in space' and the deployment of anti-satellite weapons would is disheartening (Krepon and Clary, 2003). Anyone with a cellphone, computer, or GPS system is vitally dependent on space-based communication and navigation systems, and this is equally true for the peoples of undeveloped countries as it is for those from the industrialized north. We already face numerous challenges to the continued use of outer space for civilian purposes, including dwindling

[5] To cite one of many examples, the physicist and long-time Pugwash member John Holdren, who gave the acceptance speech for the Pugwash Conferences at the Nobel Peace Prize ceremony in Oslo in December 1995, became President Obama's chief Science Advisor in 2009.

orbital slots for satellites and an increase in space debris. Should countries begin to test anti-satellite weaponry; the problem of space debris will only become worse. Hence, the problem of deploying weapons in space, or to be used in space, is one that should be able to mobilize both the scientific community and the wider public. Working through scientific and public policy organizations, pressure could be brought to bear on national governments and international bodies such as the UN, the ITU (International Telecommunications Union), the WMO (World Meteorological Organization) and others, to ensure that outer space truly remains a common resource for mankind in the decades ahead.

In seeking to mobilize international opinion in favour of a regime banning anti-satellite weapons, a useful precedent is that of the Chemical Weapons Convention that was signed and ratified in the early 1990s. I remember attending Pugwash workshops in the late 1980s on chemical weapons that emphasized engaging the civilian chemical industry as well as government, military and arms control specialists, in order to broaden the coalition of stakeholders who could then be mobilized to support the Chemical Weapons Convention. Pugwash would invite senior officials from Monsanto, Dow, and Dupont to Pugwash workshops so that these companies would be in on the 'ground floor' in helping design an international regime for eliminating chemical weapons that they could live with, given the need to inspect civilian chemical facilities. It was very important to have these companies on board as stakeholders in the process of building the Chemical Weapons Convention, and having them on board then made it easier to gain ratification of the CWC from the U.S. and other governments (Robinson, 1993).[6]

In the same way, there are widespread commercial interests in outer space that have a fundamental stake in preventing the deployment of anti-satellite weapons and the weaponisation of space (Boutwell *et al.*, 2004). From telecommunications and navigation to weather prediction and environmental monitoring, there are extensive industry and private sector interests who should be vitally concerned with maintaining the peaceful nature of outer

[6] See also reports of the Pugwash continuing series of workshops on chemical and biological weapons, organized by Perry Robinson and Matthew Meselson, in issues of the Pugwash Newsletter from the 1980s to the present.

space. The scientific and arms control communities need to do a better job of reaching out to these companies to get them involved in designing an anti-satellite convention, much the way chemical companies were involved in the chemical weapons convention in the 1980s and 1990s.

Global Climate Change and Arctic Security

The third major issue I'll address is that of military, economic and human security in the Arctic region and how these are being affected by global climate change.

As with all issues related to climate change, there are a wide range of complex inter-related problems being raised by the shrinking of the polar ice cap and a reduction in the presence of sea-ice in the shipping lanes of the Arctic region.

First is the geographic scope of the problem. The Arctic region spans 24 time zones, encompasses about one-sixth of the world's landmass, and directly includes the territory of the two major nuclear weapons states — the US and Russia — as well as those of Canada, Iceland and the Scandinavian countries.

Second is the fundamental nature of how climate change effects are exacerbated in the Arctic region. Even allowing for uncertainties in the effects of climate change in different parts of the world, there is a solid scientific consensus that climate change will have greater effects in the Arctic. As noted by Ola Johannessen of Denmark, 'global climate change is enhanced in the northern high latitudes due to complex feedback mechanisms in the atmosphere-ocean-ice system' (Johannessen, 2008). Accordingly, many scientists are predicting that Arctic temperatures over the next 50 years will increase by approximately 3–4°C, or twice the global average. If true, this will have wide-ranging consequences for diminishing the polar ice cover and for the appearance of an ice-free Arctic ocean during the summer months, as well as for a melting of the Greenland ice cap.

There remain scientific uncertainties in the pace of climate change, of course. The Arctic climate is subject to wide variability, so greater climate change science and modeling is necessary to ascertain the likely impacts in the years ahead. What is crucially important is that climate science be given the chance to provide the information necessary for informed public policy decisions before countries in and around the Arctic rush in to claim sovereignty and begin dividing up the economic and natural resource spoils of the region.

Unfortunately, national competition for the natural resources of the Arctic region has already begun, with distressing consequences. Symbolic of this is the controversy of Russia seizing a Greenpeace International ship, the *Arctic Sunrise*, in September 2013, in the Pechora Sea near a Russian gas drilling platform. According to Greenpeace, the ship was in international waters when it was boarded and seized by Russian commandos rappelling from helicopters overhead. The Russian government claims that the Greenpeace ship posed a significant risk of causing an accident and environmental damage as it neared the Gazprom drilling platform. Whatever the truth of the matter, the outcome thus far is that 30 Greenpeace activists spent several months in a Russian jail and faced the prospect of trial in a Russian court.[7]

Well beyond this one incident of international tension over Arctic resources, there is a wide variety of environmental, political, economic, military, social and cultural dimensions of changes in the Arctic region which will demand increased international cooperation. These include:

1. Changes to the ecology, habitats and lifestyles of native flora, fauna and local inhabitants (some 4 million people, including more than 30 different indigenous peoples);
2. International competition over greater access to seabed minerals and resources (including perhaps as much as 20 percent of the world's untapped oil and natural gas reserves);
3. Increased navigation through the waters of the Arctic Ocean (the famed Northwest passage) that could transform international shipping;
4. The impact of rising sea levels around the globe resulting from the melting of the Arctic and Greenland ice caps;
5. Increased militarization of the Arctic region given the proximity of the US and Russian military forces, including nuclear weapons submarines.

[7]The captain of the *Arctic Sunrise*, Peter Willcox, was also the captain of the famous Greenpeace vessel, *Rainbow Warrior*, which was sunk by French agents with the loss one life in a New Zealand harbour in 1985 during protests over French nuclear weapons testing in the South Pacific in 1985 (Kramer, 2013).

Perceptions of Arctic seabed riches and lucrative shipping profits are such that already we find comical Chaplinesque displays of countries asserting their national sovereignty in the region, such as Russia's placing its national flag on the seabed floor and Canada sending its Defence Minister to a tiny outpost on Hans Island near Greenland, to which Denmark promptly protested.[8] While outwardly comical, however, these displays of nationalism presage an increasingly earnest competition in the region that the international community would do well to prevent before it turns into conflict.

One means for doing so would be to turn to the precedent for international scientific cooperation established in the 1959 Antarctic Treaty. While acknowledging the differences between the south and north polar regions, the premise of the 1959 treaty for using the Antarctic region 'exclusively for peaceful purposes' is an excellent starting point for strengthening international cooperation in the Arctic. Secondly, there is the Antarctic Treaty confidence-building measure of allowing observers from any Contracting Party to visit the civilian bases and research stations of other contracting parties. As with other international peace and arms control agreements, the principles of transparency and the sharing of data (all of which are fundamental to scientific discourse) are also key to promoting the understanding and trust necessary to reach agreements among those in dispute and conflict with each other.

In September 2004 a permanent Secretariat to support the Antarctic Treaty and its implementation was created in Buenos Aires, which could provide a further model for scientific cooperation for the Arctic. As noted on the website of the Antarctic Treaty, the Secretariat is tasked to support the annual Antarctic Treaty Consultative Meeting (ATCM) and the Committee for Environmental Protection (CEP); facilitate the exchange of information between the 12 states parties as required by the Treaty and the separate Environmental Protocol; collect, store, archive and make available documents of the ATCM; and provide and disseminate information about the Antarctic Treaty system and Antarctic activities.[9] All of these measures help increase transparency and the effectiveness of both national and joint international research activities in the Antarctic region while minimizing the potential for friction and conflict.

---

[8]The dispute over the 1.3 square kilometre island near Greenland has been ongoing for many years, yet has still not been fully resolved (Campbell, 2012).

[9]The website of the Secretariat for the Antarctic Treaty is http://www.ats.aq/e/about.htm.

Given the additional status of the Antarctic as a nuclear weapon-free zone (NWFZ), there are, to be sure, fundamental differences between the two polar regions in terms of the military presence of nuclear-armed countries. Not only do American and Russian nuclear weapons and attack submarines regularly patrol in Arctic waters, but the Russians maintain their largest SLBM base within the Arctic Circle on the Kola Peninsula.

In recent years, several organizations, including the Canadian Pugwash Group, have called for making the Arctic a nuclear weapon-free zone (Buckley, 2012). Such an outcome will not happen anytime soon, given the enormous difficulties of rolling back long-standing nuclear weapon and other military activities in the region. But the objective of giving the Arctic region the same non-nuclear weapon status as that of Antarctica does provide great scope for international scientific cooperation in terms of researching the monitoring and verification technologies that would be needed to help implement such a treaty.

One such example has been provided by Michael Wallace of the University of British Columbia, who notes that Canada considers the Northwest Passage to be within Canadian internal waters, while the United States and the European Union consider the passage to be an international channel (Wallace, 2007). If ultimately judged to be Canadian, the government in Ottawa could take steps to ban the surface transit of all nuclear materials, thus de facto creating a nuclear weapon-free zone (NWFZ) for international shipping in the region. To enforce such a ban, the international scientific community could help with the necessary monitoring and verification technologies.

Wallace acknowledges that even creating a ban on the transit of nuclear materials in the Northwest Passage is far removed from implementing a full-fledged nuclear-free Arctic covering SLBM submarines. Nonetheless, a useful precedent would have been established, looking forward to a time when further reductions in global nuclear arsenals make such an Arctic NWFZ possible.

Long before such a NWFZ becomes feasible, what will be needed are mechanisms for cooperative scientific and policy discourse on the wide range of important Arctic issues. A few years ago, the then current UN High Representative for Disarmament Affairs, Sergio Duarte, proposed an Arctic security regime that would head off potential competition and conflict for resources and military advantage in the region. Among a number of components of such a regime that would promote communication and confidence-building measures in the Arctic, Duarte proposed 'cooperation in the area of scientific research and the development of technology to protect the environment and to improve the living conditions of those who live in the region' (de Queiroz Duarte, 2008).

To that end, non-governmental organizations should be urging their governments to establish scientific working groups and other modalities for in-depth investigation of how various scenarios of global climate change will impact the Arctic. Some work along these lines has already begun, but much more needs doing if the Arctic is to avoid becoming a hot zone of international political, economic and military competition.

Summary

The following is a brief summary of the various means by which science diplomacy and scientific cooperation and transparency can be a positive force in global politics:

1. Individual scientists can make a difference, as the examples of Rotblat, Meselson, and al-Shahristani demonstrate;
2. The communications and information revolutions of the past 50 years have made basic scientific tools and knowledge available to anyone with a computer, iPad, or cellphone;
3. Social networking now makes possible the creation of more narrowly focused and effective interest groups across much wider areas of the globe;
4. The ability of various scientific organizations, whether governmental (US National Academy of Sciences) or non-governmental (the Pugwash Conferences on Science and World Affairs), to offer independent scientific analyses and critiques of major policy issues remains an important component of international cooperation;
5. The major role played by science and technology advances in many of the current challenges to international peace and security — be they in the form of the possible weaponization of outer space or national competition for Arctic resources stemming from global climate change — is such that the scientific community is well-positioned to use its long history of international cooperation to find cooperative solutions to these problems.

In sum, the traditions of objective assessment, transparency and cooperation inherent in the international scientific community, coupled with new communications and information technologies that allow greater public involvement in these debates, will remain a positive force for finding new and creative solutions to enduring problems of human and international security.

References

Bethe, H., Boutwell J. and Garwin R.L. (1986) 'BMD Technologies and Concepts in the 1980s', in Long, F.A., Hafner, D. and Boutwell, J. (eds.) *Weapons in Space* volume 1, New York: W.W. Norton, pp. 53–72.

Boutwell, J., Hitchens T. and Moltz J.C. (eds.) (2004) 'Ensuring Security in Space: Enhancing Stakeholder Cooperation', *AstroPolitics: The International Journal of Space, Power and Policy*, 2(2): 99–106.

Buckley, A. (2012) 'An Arctic Nuclear Weapon-Free Zone: Needed Now', *Cadmus*, 5 November, [Online], Available: http://www.cadmusjournal.org/node/259.

Campbell, J. (2012) 'Hans Island deal downplayed; "Several options on table", Danish ambassador says', *The Ottawa Citizen*, 31 May, [Online], Available: http://www2.canada.com/topics/news/national/story.html?id=6475505.

Chayes, A and Wiesner, J.B. (eds.) (1969), *ABM: An Evaluation of the Decision to Deploy an Anti-ballistic Missile System*, London: Macdonald and Co.

de Queiroz Duarte, S. (2008) 'Keynote Address', in *Arctic Security Conference Report*, Vancouver, BC: Simon Fraser University School for International Studies, Available: http://nautilus.org/wp-content/uploads/2011/12/Dhanapala-et-al-Arctic_Security_Conference-NWFZ-2008.pdf, pp. 6–8.

Evangelista, M. (2002) *Unarmed Forces: The Transnational Movement to End the Cold War*. Ithaca, NY: Cornell University Press.

Hafner, D.L. (1986) 'Assessing the President's Vision: The Fletcher, Miller and Hoffman Panels', in Long, F.A., Hafner, D. and Boutwell, J. (eds.) *Weapons in Space,* volume 1, New York: W.W. Norton, pp. 91–126.

Johannessen, O.M. (2008) 'Arctic Climate: Present and Future Perspective', in *Arctic Security Conference Report,* Vancouver, BC: Simon Fraser University School for International Studies, Available: http://nautilus.org/wp-content/uploads/2011/12/Dhanapala-et-al-Arctic_Security_Conference-NWFZ-2008.pdf, pp. 10–11.

Kramer, A.E. (2013) 'Captain of Seized Greenpeace Ship Speaks from Russia', *New York Times*, 23 November, [Online], Available: http://www.nytimes.com/2013/11/24/world/europe/captain-of-seized-greenpeace-ship-speaks-from-russia.html.

Krepon, M. and Clary, C. (2003) *Space Assurance or Space Dominance? The Case Against Weaponising Space,* Washington, DC: Henry L. Stimson Centre, Available: http://www.stimson.org/images/uploads/research-pdfs/spacebook.pdf.

Pilkey, O.H., Pilkey K.C. and Fraser M.E. (2011) *Global Climate Change: A Primer,* Durham, NC: Duke University Press.

Robinson, J.P. (1993) 'The 1993 Chemical Weapons Convention', *Bulletin of Arms Control,* London: Centre for Defence Studies, Available: http://www.fas.harvard.edu/~hsp/chemical.html.

Rotblat, J. (1972) *Scientists and the Quest for Peace: A History of the Pugwash Conferences* Cambridge, MA: MIT Press.

Salama, S. and Hunter C. (2005) 'Leading Iraqi Nuclear Scientist, Once Imprisoned, Elected to Prominent Post', *CNS Research Story,* 7 June, [Online] Available: http://cns.miis.edu/stories/050615.htm.

Stares, P. (1985) *The Militarization of Space: U.S. Policy 1945–1984,* Ithaca, NY: Cornell University Press.

Tucker, J. (2001) 'The "Yellow Rain" Controversy — Lessons for Arms Control', *The Nonproliferation Review,* 8(1): 25–42. DOI: 10.1080/10736700108436836.

Wallace, M. (2008) 'Nuclear Weapon Free Zone in the Arctic? A Step by Step Approach to Overcoming the Obstacle', in *Arctic Security Conference Report,* Vancouver, BC: Simon Fraser University School for International Studies, Available: http://nautilus.org/wp-content/uploads/2011/12/Dhanapala-et-al-Arctic_Security_Conference-NWFZ-2008.pdf, pp. 33–37.

Yonas, G. (1986) 'The Strategic Defense Initiative', in Long, F.A., Hafner, D. and Boutwell, J. (eds.) *Weapons in Space,* New York: W.W. Norton, pp. 73–90.

Chapter 11

Global Health Research Diplomacy

Edison T. Liu

Diplomacy is defined as 'the conduct of the relations of one state with another by peaceful means'[1] or the 'art of dealing with people in a sensitive and effective way'.[2] In both definitions, the key word is relationship with the fundamental action being to build relationships. We use different vehicles to enhance this formal relationship building. For example, sports diplomacy engages competitive sports and cultural diplomacy uses music and the arts to create nodes of interaction. Both forms of diplomacy are language independent, and derived from basic human drives: the first is to compete, and the second is to express. Sports connect people through friendly competition, and the arts link people through nonverbal communication. It is the universality of these activities that makes them so powerfully transcendent over the limitations of language and the constraints of race and ethnicity.

Science is another vehicle for diplomacy deriving its attraction from the basic human need to know and to build. Since science is viewed as neutral and objective, it carries with it the impression of being apolitical — a useful characteristic in diplomatic discussions. Within science diplomacy, the use of health research as a platform has some unique aspects: since it improves human lives, everyone can benefit from any of its successes and, unlike the output of a conference of mathematicians, everyone can understand the impact of the scientific goals and results.

I will explore this niche area of diplomacy that uses global health research as the medium for interaction as an especially potent use of science towards

[1] The Free Dictionary, February 2014, http://encyclopedia2.thefreedictionary.com/Diplomatic+relations.

[2] Oxford Dictionary, February 2014, http://www.oxforddictionaries.com/us/definition/american_english/diplomacy?q=diplomacy.

bringing people together. Like all aspects of technologies, misapplication can also have negative consequences and in health research, cultural sensibilities play a large role in perceptions including the fear of exploitation and concerns over religious propriety. The heightened impact and the increased sensitivities suggest that diplomacy using health research as the platform should be viewed with special consideration.

I will use three examples of projects that I have engaged in that highlight the special features of this form of science diplomacy: epidemic research; clinical cancer research; and population genetics research.

Global Epidemic Response

In February 2003, World Health Organization (WHO) officers were alerted by the staff at the Hanoi French Hospital of a strange respiratory disease that proved to be fatal to a traveller from Hong Kong (Oxford *et al.*, 2003). Later, it was discovered that this traveller stayed at the same hotel at the same time as others afflicted with the same disorder. On 12 March 2003, the WHO in Geneva alerted the world about a new disease: Severe Acute Respiratory Syndrome (SARS). Because of the WHO Global Outbreak Alert and Response Network, and Global Public Health Intelligence Network, this new syndrome came to light. In the ensuing months, the WHO played a critical role as broker of collaboration and communications, and the source of central information about the world wide epidemic of SARS.

The uniqueness of this epidemic was how scientific and government health groups throughout the world cooperated to resolve this epidemic in record time. They assisted the US Centres for Disease Control and Prevention (CDC) to identify a corona virus as the culprit, facilitated the hand off of viruses and reagents so that an early diagnostic tool was constructed, which led to the full sequencing of SARS viral genome on 12 April. In May, Dutch researchers, using animal models, proved conclusively that the SARS virus caused the exact clinical disease. Also in May, the Genome Institute of Singapore published the sequence of a number of isolates and with these data constructed the SARS diagnostic that was licensed by Roche for world-wide distribution. The WHO cajoled, encouraged, and advised all parties to cooperate towards the common goal of identifying and controlling the disease (Cheng *et al.*, 2007; Heymann and Rodier, 2004) (Figure 1).

1997: WHO establishes Global Outbreak Alert and Response Network and Global Public Health Intelligence Network

Nov 2002: Retrospective data showing origin of SARS in Guangzhou

Feb 2003: Through a Chinese doctor staying in HK hotel, virus passed to Singapore, Vietnam, Toronto, and Hong Kong.

Feb 26, 2003: WHO alerted by staff at Hanoi **French Hospital of a strange respiratory disease in infected traveller from HK.**

March 12, 2003: First WHO (Geneva**) alert of a new disease SARS**

March 17, 2003: First WHO sponsored video conference of clinicians and scientists

March 22, 2003: Corona virus identified as the likely pathogen by Hong Kong **scientists**

April 12, 2003: Canadian **scientists sequence SARS virus**

May 15, 2003: Dutch **researchers prove SARS corona virus causes disease**

May 24, 2003: Singaporean **scientists show genetic variation in virus used to make robust diagnostic**

Figure 1. The timeline of events in the SARS virus epidemic.

As a player in the process, I was struck not only by the coordination of information flows, but the peer pressure that the collective interaction placed on each member. Any attempts to sequester information were immediately rebuffed; any sense of complacency was met with group criticism. Whereas, in the past, countries may have hidden their mishandling of epidemics, the speed of response and the transparency of the reporting left no malfeasance undetected.

Remarkably, the time from ascertainment of the new clinical disorder to the sequencing of the genome of the offending agent and the closure of the epidemic was only six months. This dramatic shortening of the time of onset of the epidemic to the determination of the etiological agent was due to a few key factors. Clearly, the availability of new and powerful sequencing technologies played a major role but, unquestionably, the level of international coordination and collaboration was a critical factor. The sharing of samples across governmental jurisdictions, the free access to the critical DNA data, and the coordinated information flow — including best practices — accelerated the recovery.

The SARS crisis was an example of health research diplomacy at its most effective best. The individual national scientific groups were the engines for innovation and discovery but voluntarily permitted coordination by WHO.

The important characteristics of success were willing collaboration, establishment of interaction networks, and the emplacement of lasting infrastructure and capabilities. Participants were left better than they were before the crisis.

A clear sign of this was the subsequent response in Singapore and globally around the threatened H1N1 influenza pandemic. Because of the SARS crisis, Singapore had developed a crisis management structure to coordinate governmental response and to minimize bureaucratic territorial infighting. Our scientists at the Genome Institute of Singapore were part of the management process to provide the critical research support. When H1N1 was first discovered in Mexico, to be a potentially deadly virus, the government activated the response network and our institute developed, within two weeks and through computational approaches, a H1N1 re-sequencing array that we knew would be faster, cheaper, and indeed better than next generation sequencing for the specific monitoring of H1N1 strains. This was linked to a web based reporting system established by the Bioinformatics Institute (BII) to immediately upload sequence data from isolates provided by the Ministry of Health physicians, and to provide a computational rendering of the sequence data to assess the origins of the virus, and to determine the probability of resistance to antivirals (Hurt *et al.*, 2011; Lee *et al.*, 2011). This system was established in a short period of time and was used throughout the epidemic by policy makers to monitor the genetic mutations of the flu viral isolates entering Singapore that could lead to drug resistance. In sum, the global health diplomacy at the time of SARS strengthened Singapore's infectious disease network and left us with a more robust pandemic response infrastructure.

Contrast this experience with another episode in the history of epidemics. In 1894, an outbreak of the bubonic plague that claimed over 40,000 lives captured the attention of the world's burgeoning biomedical research community. Armed with recent knowledge of microbiology, especially in pathogen isolation, Alexandre Yersin of France and Shibasaburo Kitasato of Japan both arrived in Hong Kong with their party of scientific assistants. Their primary purpose was to isolate and characterize the bacterium responsible for the plague, of course, with the hope that they could then devise appropriate therapies. Working tirelessly, both groups succeeded in isolating the causative agent which ultimately was renamed in 1967 as *Yersinia pestis.*

Both scientific groups stayed in Hong Kong simply long enough to characterize the first isolates and then left. None trained local physicians or left enduring infrastructure in China, though Yersin spent the rest of his life in Saigon at the Pasteur Institute continuing his studies in tropical diseases. The sad irony was that later during World War II, the Japanese military used *Yersinia pestis* as a biological warfare agent against Chinese populations. Though the work of Yersin and Kitasato was ground breaking, their involvement in China was one of extraction rather than diplomacy. None of the tenants of scientific diplomacy, parity of respect, building relationships, and leaving a positive legacy, were honoured during the plague of 1894 (Bibel and Chen, 1976).

Like all aspects of diplomacy, there is a formality to the process that often culminates in a concrete output. It may be a truce, a communiqué, or a treaty that sets in motion processes or programmes leading to an expected outcome. Research diplomacy uses the same tools. In the Human Genome Project, the US, UK, China, Germany, France, and Japan were willing participants, each contributing its own national resources to this international effort. The process was the development of approaches and technologies that could sequence and assemble the human genome, and the outcome was the complete sequence embodied in a high profile publication. At the beginning of this initiative, China was the new entrant in the field and admittedly had the least developed capabilities. Yet through this 10 year interaction, China now has the most powerful genomic facilities in the world (Cyranoski, 2010). This is another clear example of science diplomacy leaving a positive legacy in its wake.

Health diplomacy can also be a vehicle for conflict resolution. In 1999, I was involved in the establishment of the All-Ireland-NCI Cancer Consortium,[3] which was ostensibly set up to enhance clinical cancer research in the Republic of Ireland and Northern Ireland by engaging the expertise of the US National Cancer Institute. But the historical and political backdrop became even more important. After decades of internecine violence, factions in Northern Ireland — with the help of leadership in the UK and the Republic of Ireland — finally arrived at a peace accord. All parties were seeking projects that would allow conflict communities to willingly work

[3] For the official website, see http://www.allirelandnci.com/index.asp.

together. At the signing of the agreement at Stormont Castle, a young lady approached me and thanked me for helping establish this consortium. She added that since she was a Sinn Fein member, she would have never set foot in that icon of Unionist Northern Ireland, Stormont Castle, were it not for an initiative that sought to cure cancer. The commonality of human suffering, in this case from cancer, rallied individuals from either side of the divide to a common cause. To date, it remains one of the most successful international relations projects for the NCI (Johnston and Daly, 2001). This was an example of science diplomacy that had its greatest role as a bridge between politically distant communities.

The Human Genome Organization[4] is the oldest organization dedicated to enhancing the sequencing and the study of the human genome. As genomic sciences accelerated after the first sequence draft of the human genome in 2001, members from the Asia Pacific region wanted to come together to work on a collaborative project that was not dependent on the goodwill and leadership of the west. This was not a rejection of our friends in the Americas and Europe, rather, it was a declaration of an aspiration for scientific self-reliance, since most key international scientific arrangements in which Asian entities engaged were led by a western institution, and none were conceived, executed, and funded by Asian entities alone. Moreover, we understood that for human genetics to advance, the precise diversity of the Asian gene pool, which accounts for nearly half of humanity, should be assessed.

There were almost insurmountable challenges: in 2002, there were dramatic differences in scientific and/or financial capabilities amongst the Asian nations. So any Pan-Asian project that was inclusive would have to deal with a range of genomic and genetic infrastructures, ranging from world-class in Japan; to aspirational but nascent (and sometimes barely operational) in Singapore, Korea, Taiwan, India, and China; to wanting in Indonesia, the Philippines, and Thailand. There was also a significant disparity in population sizes (e.g. Singapore vs. China or India). We had to deal with government regulations that prohibited the delivery of human DNA outside a country without the expressed approval of national bodies. There

---

[4]See the official website, http://www.hugo-international.org/, for more details, specifically the 'About Us' page for the year it was conceived.

was the problem of funding: most Asian countries do not have a policy or history of sending precious internal scientific funds to an outside agency for redistribution. Finally, there were historical animosities and national rivalries that limited inter-Asian cooperation: many countries had difficulty ceding leadership to Japan because of the memories of their imperialist and colonial past, Malaysia was competitive with Singapore, and Taiwan's inclusion was problematic for China. Moreover, many of the investigational entities in Asia were funded by ministries of trade and industry whose export-minded strategies viewed other Asian nations as competitors. The idea of helping a competitor in a potential biotech developmental sector would be a difficult sell.

The first attempts as assembling a Pan-Asian Genome Project failed because it could not bypass these problems. The initial discussions in 2002 imagined a project whereby DNA samples would be sent to Japan for genotyping, and that the study would focus on comparing the genetic variations in individuals with disease and those without disease. This case-control design presupposed that each country would be able to identify and recruit patients and normal volunteers for genetic assessment, which is expensive and requires time to set up. Moreover, for such a study to be effective, many thousands of cases need to be recruited — numbers count.

Between 2002 and 2004, a small group of us conceived of a plan that, ultimately, was successful. The study we proposed was to assess the genetic diversity (and relatedness) of Asian populations which would establish the first database of Asian single nucleotide polymorphism (SNP) diversity that would be useful and open to all. The study design was to identify DNA samples from individuals representing the greatest ethnic diversity from our collaborators, and, using the then newly developed standardized 50,000 SNP arrays from Affymetrix, to experimentally determine our Asian genetic diversity. This design avoided any need for costly disease ascertainment and subject accrual, and immediately raised the importance of countries harbouring the greatest ethnic diversity like the Philippines, Thailand, and Indonesia in this collaboration even though they had the least developed infrastructure for this type of research. Our parallel developmental goals were to enhance Pan-Asian scientific cooperation and to establish national infrastructures that would enable future Asian-centric collaborative scientific networks around population genetics.

To circumvent the problem of disparate resources amongst our collaborators, we used the principles of the host-guest relationship familiar to all of us in Asia where hosts are generous and humble, and guests, when honoured, are polite and non-demanding. In this construct, resource-enabled countries act as hosts by providing genotyping costs, and training, and 'guest' country scientists bring their national DNA samples to the host institutions for genotyping. Because the guests keep control of their DNA samples throughout the analyses and take them home afterwards, there is no loss of custody. Finally, the data generated were open to all collaborators, but the analysis was performed as a collaborative group so that we could learn from each other. This arrangement was structured to build trust (through deep interactions), to enhance capabilities (through training), and to provide parity of respect (host-guest relationships, and that the most crucial participants were the less-enabled countries with the greatest genetic diversity).

The final study engaged 11 countries and 30 institutions, 90 investigators, studying 1,903 individuals from 73 distinct ethnic populations. The impact of the results were surprising even to the participants and led us to redraw the map of human migration into Asia (HUGO Pan-Asian SNP Consortium, 2009). These data were subsequently used by our colleagues for their own national genetic studies (Ngamphiw *et al.*, 2011; Hatin *et al.*, 2011; Xu *et al.*, 2010). In the end, we achieved not only scientific success, but also made social and political progress.

Each of these historical vignettes are examples of good, and to failed, science diplomacy. Taken together, these stories frame the criteria for success in such endeavours. It starts with all participants working together with equal respect, with a goal of building relationships, and leaving structures, organizations, or a base of knowledge that benefits everyone (Table 1).

Table 1. Components for successful diplomacy in science.

	Science diplomacy, criteria for success
1	Parity of Respect: Are the participants viewed as equal partners?
2	Building Relationships: Are personal networks formed?
3	Leaving a Positive Legacy: Are the participants better off or more enabled than before?

The New Hanseatic League in a Technologic World

I defined science diplomacy as using science to increase relations between peoples and nations. But diplomacy is also a non-military way to increase one's influence — a manifestation of soft power. For small nations like Singapore and New Zealand, diplomacy is often used to leverage influence beyond their size and resources. These nations, with a range of populations from 4–7 million, need to function differently from larger countries because of their limited depth in science and limited financial resources, which demands that they have a focussed policy to achieve international competitiveness. The advantages of a small country are that they usually have a highly qualified work force with niche expertise, and that there is a close relationship between decision makers, academics, and the technocratic work force, which allows for quick response times and rapid implementation.

In an increasingly competitive world where everyone struggles on a global scale, small countries often pair up with larger countries only to find that their interests are not a priority to their partners. However, there may be another strategy drawn from the past that allows for small nations to be

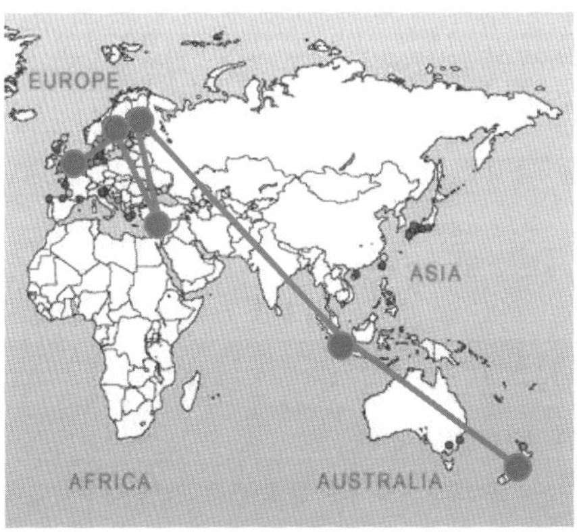

Technology "Hanseatic" League:
Confederation of small nations with complementary critical mass
Prearranged IP sharing structure
Reduction of data sharing barriers
Differentiated and coordinated research activities
Projects with definable endpoint or product
Shared web-based teaching modules

Figure 2. Possible arrangement of a 'Hanseatic' League of small nations with technology capabilities and strengths.

globally competitive in science and technology. In the 12th–15th centuries, a confederation of around 60 European city states assembled as the Hanseatic League to enhance their mercantile competitiveness. They banded together for mutual protection, to reduce trade barriers, and to differentiate their activities so as to be mutually beneficial. In this manner, they dominated trade in the Baltic region. Perhaps, it is an opportune time to form a modern technology-based 'Hanseatic League' for similar mutual benefit.

Other small nations such as Finland, Israel, Ireland, and Norway may want to join to pre-arrange intellectual property sharing arrangements, reduce barriers to data-integration and sharing, co-ordinate research activities that take advantage of complementary skills, and local critical mass towards definable endpoints or products (Figure 2).

Diplomacy is one of the most significant manifestations of an advanced civilization. Weaponry has progressed to a point where a full frontal conflict would have disastrous consequences, so diplomatic solutions become paramount for conflict resolution and therefore for human survival. With science and medicine becoming so prominent in our lives, the power of this common experience in discovery and research may very well bring us all closer together.

References

Bibel, D.J. and Chen T.H. (1976) 'Diagnosis of Plague: An Analysis of the Yersin-Kitasato Controversy', *Bacteriological Reviews*, 40(3): 633–651. Available: http://mmbr.asm.org/content/40/3/633.full.pdf.

Cheng, V.C., Lau S.K., Woo P.C. and Yuen K.Y. (2007) 'Severe acute respiratory syndrome coronavirus as an agent of emerging and re-emerging infection', *Clinical Microbiological Review*, 20(4): 660–94. DOI: 10.1128/CMR.00023-07.

Cyranoski, D. (2010) 'Chinese Bioscience: The sequence factory', *Nature*, 464(7285): 22–4.

Hatin, W.I., Nur-Shafawati A.R., Zahri M.K., Xu S., Jin L., Tan S.G., Rizman-Idid M., Zilfalil B.A. and HUGO Pan-Asian SNP Consortium (2011) 'Population genetic structure of peninsular Malaysia Malay sub-ethnic groups', *PLoS One*, 6(4): e18312.

Heymann, D.L. and Rodier G. (2004) 'SARS: A Global Response to an International Threat', *Brown Journal of World Affairs*, 10(2): 185–197.

HUGO Pan-Asian SNP Consortium (2009) 'Mapping human genetic diversity in Asia'. *Science*, 326(5959): 1541–5. DOI: 10.1126/science.1177074.

Hurt, A.C., Lee R.T., Leang S.K., Cui L., Deng Y.M., Phuah S.P., Caldwell N., Freeman K., Komadina N., Smith D., Speers D., Kelso A., Lin R.T., Maurer-Stroh S., and Barr I.G. (2011) 'Increased detection in Australia and Singapore of a novel influenza A(H1N1) 2009 variant with reduced oseltamivir and zanamivir sensitivity due to a S247N neuraminidase mutation', *Euro Surveillance*, 16(23): pii=19884. Available: http://www.eurosurveillance.org/ViewArticle.aspx?ArticleId=19884.

Johnston, P.G. and Daly P.A. (2001) 'The NCI-Ireland consortium: a unique international partnership in cancer care', *Oncologist*, 6(5): 453–8.

Lee, V.J., Yap J., Maurer-Stroh S., Lee R.T., Eisenhaber F., Tay J.K., Ting P.J., Loh J.P., Wong C.W., Tan B.H., Koay E.S., Kelly P.M. and Hibberd M.L. (2011) 'Investigation of causes of oseltamivir chemoprophylaxis failures during influenza A (H1N1-2009) outbreaks', *Journal of Clinical Virology*, 50(2): 104–8. DOI: 10.1016/j.jcv.2010.10.004.

Ngamphiw, C., Assawamakin A., Xu S., Shaw P.J., Yang J.O., Ghang H., Bhak J., Liu E., Tongsima S. and HUGO Pan-Asian SNP Consortium (2011) 'PanSNPdb: The Pan-Asian SNP Genotyping Database', *PLoS One*, 6(6): e21451.

Oxford, J.S., Bossuyt S. and Lambkin R.A. (2003) 'New infectious disease challenge: Urbani severe acute respiratory syndrome (SARS) associated coronavirus', *Immunology*, 109(3): 326–328.

Xu, S., Kangwanpong D., Seielstad M., Srikummool M., Kampuansai J., Jin L. and HUGO Pan-Asian SNP Consortium (2010) 'Genetic evidence supports linguistic affinity of Mlabri — A hunter-gatherer group in Thailand', *BMC Genetics*, 11: 18.

CHAPTER 12

Science, Diplomacy and Trade: A View from a Small OECD Agricultural Economy

Stephen L. Goldson and Peter D. Gluckman

Introduction

Small countries have a set of imperatives that are different from that of large countries. Their policy frameworks must take scale into account and a number of factors including the relative size of their economies make them more vulnerable to offshore events. At the same time their small-scale allows them to be more prompt in responding to change and this confers resilience. However, their voices in the changing multi-national architecture can be difficult to hear. This latter point is particularly important in the trade arena where countries such as New Zealand rely heavily on the establishment and maintenance of a rules-based open trading system (e.g. MPI, 2013). New Zealand has the further challenge of being a very open democracy but one where its geographical position in the world and the size of its economy means that it must constantly strive to maintain its interests and relevance in the global arena. New Zealand, with a population of just over four million, must be particularly strategic in how it deploys its limited diplomatic resources and, in turn, give careful consideration to how these intersect with science. By way of example, this chapter considers the diplomatic perspective on the apparently diverse issues of biosecurity and food safety on the one hand and pastoral greenhouse gas emissions primarily from ruminant-associated methane production on the other.

As a trading nation sustainable primary industries are critical to New Zealand's economic well-being and in recent years, they have increased in their importance such that they now contribute to more than 65 percent of New Zealand's merchandise exports (e.g. Anon, 2009). Enormous attention is therefore being paid to the need for its land-based sectors to maintain their reputation as reliable providers of high quality and safe food. Any issues relating to negative market perception is of paramount importance, as illustrated by the flurry of concerns over the recent episode involving infant formula possibly contaminated by *Clostridia* (the bacteria that produces the botulinum toxin). A second issue is the need for strenuous science-based biosecurity efforts to prevent pests and diseases from crossing the border, combined with appropriate biological control of pests (wherever possible) in both its managed and indigenous ecosystems as a way to minimise the use of pesticides. Integral to the marketing of food is the need to show that the ecosystems are demonstrably clean and sustainable in the face of international scrutiny of the claims that New Zealand makes about its environmental purity. Exactly the same requirement applies to protecting the country's very large tourism industry.

Based on such considerations, this chapter comprises three parts. The first covers New Zealand's need for biosecurity, food safety and trade, while the second refers to the expedient for the country to contribute to research on the global abatement of greenhouse gas emissions from agriculture. The third part pulls these ideas together by discussing and illustrating how science has been essential to informing and empowering some of the country's diplomatic interests and initiatives.

Biosecurity, Food Safety and Trade

As already mentioned, New Zealand's primary industries are essential to its economic well-being. Tourism also provides about NZ$10 billion of external earnings (Wilson and Riley, 2012) and beyond the wilderness landscapes and national parks, tourism also depends on the relatively pristine productive landscapes and increasingly, high quality wine and food experiences. Yet maintaining these environmental qualities is not particularly easy. New Zealand's trade and tourism are growing at a very high rate, both having nearly doubled over the last 10 years (Ministry of Business Innovation and Employment, 2013) and the markets are diversifying all of the time. As a result of such traffic, New Zealand is wide open to invasion by unwanted and destructive exotic organisms. Those who visit that go trout fishing do not want to be slithering around on the invasive diatom didymo (*Didymosphenia geminata*) that so negatively affects stream life (e.g. Beville *et al.*, 2012).

Furthermore, the New Zealand pastoral farming ecosystem represents a partial transplant of European farming systems comprising pastures of less biodiversity than the original habitats (Goldson *et al.*, 2014); many being little more than a mixture of imported ryegrass and nitrogen-fixing clover. Such a simplified environment, is very susceptible to invasive plants and animals that reach New Zealand. This is because there are few, if any, natural enemies, few competing species and many unfilled ecological niches. As a result, trying to do something about exotic weed, pest and disease invasion has been the focus of much of New Zealand's science, such as in the area of biological control of pasture pests like the clover root weevil (*Sitona obsoletus*) (Goldson *et al.*, 2001). The same considerations apply to the other primary sectors such as horticulture where for example, the ravages of the bacterial disease *Psa* (*Pseudomonas syringae* pv. *actinidiae*) has done so much damage in kiwifruit orchards (Greer and Saunders, 2012).

For very different reasons, New Zealand's much-admired but very fragile indigenous ecosystems are also open to invasion. In contrast to the pastoral ecosystem, here there is much biodiversity, but this has evolved over 80 million years of geographical isolation and, as a result, its species are often very poorly adapted to deal with invaders. This general defencelessness is epitomised by flightless and often nocturnal birds that are immediately vulnerable to aggressive invasive mammalian predator species such as stoats, cats, dogs, weasels and humans. Indeed, it is estimated that 40 percent of the country's native avifauna has become extinct (Worthy and Holdaway, 2002) since humans first arrived approximately 700 years ago (Wilmshurst *et al.*, 2008). Problematically, New Zealand's indigenous animal and plant populations are hugely important in terms of international biodiversity. For example, 90 percent of all of the insect species found in New Zealand are endemic (McGlone, 2006). In 1992 New Zealand signed up to the international *Convention on Biological Diversity* (Swanson, 1997) and has therefore signalled its intention to preserve its unique indigenous biodiversity.

Thus, biosecurity is of paramount importance to New Zealand because of its requirement to protect New Zealand's productive economy and its biodiversity. Indeed, it would be fair to assert that there is a higher degree of public concern in New Zealand about biosecurity than found elsewhere. But with this comes a paradox. While New Zealand has to be deeply concerned about biosecurity, it is also very dependent on free trade. Given that the country has a small internal economy, there is a huge dependency on exports; for example 90 percent of its milk and milk products is shipped to overseas markets (Anon, 2010). Loss of access to markets through diplomatic mishaps, tariff barriers, non-tariff barriers, currency fluctuations and of course, food safety scares can be catastrophic. For this reason New Zealand unswervingly supports World Trade Organization (WTO)-based free trade agreements, regional trade initiatives etc., yet at the same time it must get its trading partners to adhere to potentially misunderstood biosecurity compliance requirements and to what may be seen to be inconvenient, biosecurity–related red tape.

Domestically and internationally serious arguments can break out about biosecurity requirements being wrongly, ineptly or cynically applied; these often occur along with thinly-veiled accusations of non-tariff barriers (e.g. Gluckman *et al.*, 2012). Two seemingly perennial examples concern the importation into New Zealand of uncooked pig meat (Watkins, 2013) and honey products (Oughton *et al.*, 2009). The New Zealand domestic industries are saying that their proclaimed resistance to such imports is about protecting the disease-free nature of their enterprise and New Zealand. Conversely, overseas exporters claim that it is all about New Zealand industries not wanting competition from more efficient producers elsewhere and that the risks really are minimal. The converse argument occurred with regard to the export of New Zealand apples to Australia, which was precluded for decades because of New Zealand's fire-blight 'problem' (Knight, 2005) (more on that later). These sorts of trade/biosecurity issues can lead to long and unproductive disputes. If trade is to be encouraged, such disagreements ultimately have to be resolved using agreed-upon scientific and technical guidelines as laid out in negotiated international agreements.

Such circumstances highlight an important point. Globalisation can work in favour of a small agricultural producer like New Zealand providing that the country's science capability is up to the task of supporting those initiatives made possible through the increasing interconnectedness of the international environment. By the same token, any slippage of standards can be disastrous.

Worldwide there is growing health and environmental demand for ensured supply chain integrity. This can cut two ways. It can be a valuable marketing tool with which to try to secure an edge in high-end consumer markets, or conversely, expectations can be applied over-zealously in local regulatory systems within markets to the extent that it creates non-tariff trade barriers.

The key for New Zealand is to know more than anyone else about how biosecurity and food safety claims are mounted and the metrics behind them. In this way, the country can benefit from the quality of its own claims and yet protect its interests through good biosecurity measures without risking countervailing non-tariff barriers. Of course, assurance must be firmly based on objective and scientific verification, rather than advocacy and promotional approaches.

As discussed below it can be seen that where biosecurity, trade and food safety intersect, there is requirement for, and indeed a long history of, intense interaction between New Zealand's scientific and diplomatic communities.

Greenhouse Gas Production and Methane

The second consideration relates to the risks associated with greenhouse gas emissions produced from pastoral agriculture. Just less than 50 percent of New Zealand's greenhouse gases come from agricultural ruminants (Clark *et al.*, 2011). For an advanced economy this proportion is uniquely high and reflects the relatively large size of the pastoral sector and the country's renewable energy production (hydroelectric and geothermal). It is worth noting that although New Zealand's actual contribution to global greenhouse gas production is miniscule at 0.17 percent of the world total, the country is however, a relatively heavy greenhouse gas emitter per unit of GDP, coming second only to Australia among the OECD jurisdictions.

These agricultural emissions comprise mainly nitrous oxide and methane; the latter is produced by 31 million sheep and 10.3 million cattle (Anon, 2013). This is not trivial; every day each dairy cow produces the equivalent of 370 litres of methane per day (Lassey *et al.*, 1997) and therefore New Zealand is in this sense a huge animal bioreactor that largely turns pastoral forage into meat or milk via 700 million litres of microbe-filled rumen fluid. Methane is a particularly potent greenhouse gas as, on a molecule-to-molecule basis, it provides 25 times more forcing of heat retention than carbon dioxide. Each New Zealander is responsible for five times more methane production per capita than the global average and, thus, New Zealand has considerable interest in considering and mitigating ruminant production of methane. By doing so New Zealand is fulfilling its requirement to act as a good global citizen while maintaining its ability to export ruminant-based products and minimise criticism from its markets.

Consistent with this, New Zealand has responded directly to the challenge that it should seek to become a leading exporter of technology that mitigates greenhouse gas emissions from agriculture. One of the few positive outcomes of international discourse on climate change in 2009 was the development of the initiative proposed by New Zealand for a global research effort to reduce greenhouse gas emissions associated with agriculture. New Zealand led the discussion and provided the Secretariat for the *Global Research Alliance on Agricultural Greenhouse Gases* (e.g. Shafer *et al.*, 2011; Gluckman *et al.*, 2012) that now involves 33 nations. Its membership includes all of the major economies and agricultural producers. Therefore there is now very active collaborative research underway both in developed and developing countries. Notably, despite its small size, New Zealand is funding to a significant level this international greenhouse gas research effort, as it relates to temperate pastoral agriculture.

The Role of Science in New Zealand Diplomacy

The two areas discussed above point to how New Zealand's diplomatic strategy comes down to ensuring the country's relevance in the international arena by protecting its interests in an environment of continuingly changing global political and trade imperatives. Ultimately this is about protecting its trade in its broadest sense and this needs high levels of diplomatic vigilance. In short, New Zealand's international relations are in no small part trade-driven as well as being linked to a strong societal sense of values and place. For this country, even apparently non-trade decisions need to be evaluated through the lens of how things may affect its trading relationships.

As already mentioned the two areas of New Zealand science endeavour that relate to diplomacy are the biosecurity/trade nexus and the abatement of ruminant-derived greenhouse gases (Gluckman *et al.*, 2012). Both of these are clear examples of where New Zealand has had to be resourceful to achieve what is required to protect its interests. It is in these areas (and others) that the role of science has been essential to the broad diplomatic effort. Indeed, armed with suitable scientific insight and argument, New Zealand diplomats and officials have been (and are) particularly adroit when it comes to international trade agreements. In such areas, New Zealand diplomacy is exceptionally watchful, articulate and well respected (Gluckman *et al.*, 2012). The quality of the science itself and the high level of scienctific understanding amongst our trade negotiators has been central to extant trade-related agreements and conventions. Developing these agreements has demanded that diplomats and trade negotiators should work together, informed by science.

Biosecurity and trade

This area is exemplified by the 1952 International Plant Protection Convention (IPPC) (Schader and Unger, 2003). This international treaty aims to secure coordinated, effective action to prevent and to control the introduction and spread of pests of plants and plant products. The primary focus of this initiative is on plants and plant products moving in international trade and extends beyond the protection of cultivated plants to the protection of wild plants. It also takes into consideration both direct and indirect damage by pests, so it includes weeds. Further, the IPPC's focus is not just on plants and plant products moving in international trade: it includes any item or good that may provide a means for the introduction of plant pests into previously uninfected areas. For example, insect eggs laid on the wheels of imported cars.

Relatively recently New Zealand officials became instrumental to giving effect to the IPPC by developing detailed International Standards for Phytosanitary Measures (ISPMs) (Schader and Unger, 2003) of which there are now about 40. This collection is co-ordinated by the IPPC Secretariat which is provided by the Food and Agriculture Organization of the United Nations. The adopted standards are not regulatory instruments but serve as guidelines and recommendations to IPPC contracting parties — 179 countries, including New Zealand — in implementing phytosanitary measures. The ISPMs are also recognised as the basis for phytosanitary measures applied in trade by members of the WTO under the SPS Agreement (explained in the next paragraph). The subjects covered are quite wide ranging and include *Guidelines for Pest Risk Analysis, A Code of Conduct for The Import and Release of Exotic Biological Control Agents* and *An Export Certification System* (Kairo *et al.*, 2003).

A more recent and significant example of where New Zealand science, scientists and science-qualified officials — largely from the Ministry of Agriculture and Forestry (now the Ministry for Primary Industries) and the Ministry of Foreign Affairs and Trade — have made a very significant contribution has been in the development of the 16 pages of carefully-crafted prose that makes up the *WTO Agreement on the Application of Sanitary and Phytosanitary Measures* (known as the SPS Agreement mentioned earlier) negotiated during the Uruguay Round of the General Agreement on Tariffs and Trade in early 1995 (Scott, 2007). This was one of the final documents approved in the Uruguay Round and applies to all sanitary (relating to animals and humans) and phytosanitary (relating to plants) measures that may have a direct or indirect impact on international trade. The SPS Agreement includes a series of understandings on how SPS measures will be used by countries when they establish, revise, or apply their domestic laws and regulations. On this basis, countries have agreed to base their SPS standards on science and, as guidance for their actions, the Agreement encourages countries to use those standards set by international standard-setting organisations. In this process the SPS Agreement seeks to ensure that SPS measures are not arbitrarily or unjustifiably used to discriminate against trade of certain other members or to be used to disguise trade restrictions. The SPS Agreement accommodates the ability for countries to maintain their sovereign right to provide the level of protection they deem appropriate, but participant nations agree that this right will not be misused for protectionist purposes nor result in unnecessary trade barriers.

As already referenced in this chapter, the value of this Agreement has recently been exemplified by the case of the alleged fire-blight 'problem' where Australia refused to import New Zealand apples in order to safeguard the country against a disease that possibly it may already have had (e.g. Rodoni *et al.*, 1999). New Zealand argued that scientific evidence did not support the continuation of these imposed restrictions and opted to take the dispute to the WTO because normal diplomatic processes had failed to resolve the matter. Eventually the impasse was resolved in favour of New Zealand using the WTO dispute settlement process which ruled in New Zealand's favour based on arguments related to the level of risk and the risk management required (e.g. Westmore, 2010)

The abatement of ruminant-derived greenhouse gases

As mentioned above, the Global Research Alliance on Agricultural Greenhouse Gases is another clear and successful example of science diplomacy led by a small nation. Co-ordinated international activity in this area commenced somewhat later than the biosecurity and trade initiatives when it became widely recognised that agricultural emissions are expected to rise by about 30–40 percent above 2005 levels in line with the projected dramatic need to increase in food production by 2050 (Rosegrant and Cline, 2003. Accordingly and largely as a result of New Zealand's action, the Global Research Alliance on Agricultural Greenhouse Gases was launched in 2009 in the margins of the UN climate change conference in Copenhagen. In this, New Zealand was again at the diplomatic forefront with the International Climate Change Negotiations Minister and the Minister of Agriculture jointly hosting an inaugural three-day meeting of the Global Research Alliance in Wellington in April 2010 (Gluckman *et al.*, 2012) This comprised more than 80 senior science and policy representatives from the initial 33 nations participating (Gluckman *et al.*, 2012). New Zealand leads the livestock research group and has committed many millions of dollars both to the domestic and international research effort. These initiatives were able to be readily and immediately supported by existing New Zealand scientific expertise, particularly in rumen physiology.

While this process can be seen to relate to New Zealand's trade concerns, it has clear moral imperatives and broader diplomatic value. It demonstrates that New Zealand is indeed taking responsibility for how its primary products are being produced and is working to reduce any global impacts that may be arising from such production. It provides for north-south transfer of technology and skills to assist the many food producers in the less developed world.

Conclusion

This chapter has outlined how New Zealand's biosecurity and trade-based diplomacy is deeply rooted in and dependent on the country's science capability. New Zealand as a small but advanced country relies on the international networks that science provides to be able to best identify and incorporate knowledge derived elsewhere for its social, economic and environmental benefit. For a small country, science, economic welfare and diplomacy are very much intertwined.

Acknowledgements

The authors particularly thank Dr Ruth Frampton, Critique Limited, Christchurch, for her specific advice and insightful contribution. Thanks also to the Rt Hon Simon Upton, OECD, Paris, Dr Megan Quinlan, Imperial College, London and Dr Alan Beedle formerly Chief of Staff, Office of the New Zealand Prime Minister's Science Advisory Committee.

References

Anon (2009) *Doing business with New Zealand,* New Zealand Trade and Enterprise, https://www.nzte.govt.nz/.

Anon (2010) 'International big cheeses meet in NZ', *New Zealand Ministry for Primary Industries*, February, [Online], Available: http://www.foodsafety.govt.nz/elibrary/industry/International_Cheeses-Land_Milk.htm [10 May 2014].

Anon (2013) 'Agricultural Production Statistics: June 2013 (provisional)', *Statistics New Zealand*, [Online], Available: http://www.stats.govt.nz/browse_for_stats/industry_sectors/agriculture-horticulture-forestry/AgriculturalProduction_HOTPJun13prov/Data%20Quality.aspx [10 May 2014].

Beville, S.T., Kerr G.N. and Hughey K.F.D. (2012) 'Valuing impacts of the invasive alga *Didymosphenia geminata* on recreational angling', *Ecological economics,* 82: 1–10. DOI: 10.1016/j.ecolecon.2012.08.004.

Clark, H., Kelliher F. and Pinares-Patiño, C. (2011) 'Reducing CH4 emissions from grazing ruminants in New Zealand: challenges and opportunities', *Asian-Australian Journal of Animal Science,* 24(2): 295–302. DOI: 10.5713/ajas.2011.r.04.

Gluckman, P.D., Goldson S.L. and Beedle A.S. (2012) 'How a small country can use science diplomacy: A view from New Zealand', *Science and Diplomacy,*

25 May, [Online], Available: http://www.sciencediplomacy.org/perspective/2012/how-small-country-can-use-science-diplomacy.

Goldson, S.L., Phillips C.B., McNeill M.R., Proffitt J.R. and Cane R.P. (2001) 'Importation to New Zealand quarantine of a candidate biological control agent of clover root weevil', *New Zealand Plant Protection,* 54: 147–151. Available: http://www.nzpps.org/journal/54/nzpp_541470.pdf.

Goldson, S., Wratten S., Ferguson C., Gerard P., Barratt B., Hardwick S., McNeill M., Phillips C., Popay A., Tylianakis J. and Tomasetto F. (2014) 'If and when successful classical biological control fails', *Biological Control* 72: 76–79.

Greer, G. and Saunders C. (2012) 'The costs of Psa-V to the New Zealand kiwifruit industry and the wider community', *Agricultural Economics Research Unit, Lincoln University, New Zealand, No. 327,* Research report, [Online], Available: http://hdl.handle.net/10182/5448.

Kairo, M.T., Cock M.J. and Quinlan M.M. (2003) 'An assessment of the use of the Code of Conduct for the Import and Release of Exotic Biological Control Agents (ISPM No. 3) since its endorsement as an international standard', *Biocontrol News and Information,* 24(1): 15–27. Available: http://www.cm.colpos.mx/moodle/file.php/8/CB_Clasico/codigo_etica_importacion_EN.pdf.

Knight, J. (2005) 'Advance Australia fair? Anatomy and pathology of an 84-year trade dispute', *Journal of Public Affairs,* 5(2): 112–123. DOI: 10.1002/pa.16.

Lassey, K.R., Ulyatt M.J., Martin R.J., Walker C.F. and Shelton D.I. (1997) 'Methane emissions measured directly from grazing livestock in New Zealand', *Atmospheric Environment,* 31(18): 2905–2914. DOI: 0.1016/S1352-2310(97)00123-4

McGlone, M.S. (2006) 'Becoming New Zealanders: Immigration and the Formation of the Biota', in Allen, R.B. and Lee, W.G. (eds.) *Biological Invasions in New Zealand,* Berlin: Heidelberg Springer-Verlag, pp. 17–32.

Ministry of Business Innovation and Employment (2013) *The New Zealand Sectors Report 2013,* [Online], Available: http://www.mbie.govt.nz/pdf-library/what-we-do/business-growth-agenda/sectors-reports-series/tourism-report.pdf.

Ministry for Primary Industries (MPI) (2013) *Consultation on International and Regional Standards,* [Online], Available: http://www.biosecurity.govt.nz/biosec/pol/intl.

Oughton, M.D., Evans J. and Gough M.J. (2009) *Independent Review Panel Report on the Import Health Standard for the Importation into New Zealand of specialised bee products from Australia,* June, [Online], Available: http://www.biosecurity.govt.nz/files/regs/imports/animals/ihs-beeproic.aus-panel-report.pdf.

Rodoni, B., Kinsella M., Gardner R., Merriman P., Gillings M. and Geider K. (1999) '*Detection of* erwinia amylovora*, the causal agent of fire blight, in the Royal Botanic*

Gardens, Melbourne, Australia', *Acta Horticulturae,* 489:169–170. Available: http://www.actahort.org/books/489/489_25.htm.

Rosegrant, M.W. and Cline S.A. (2003) 'Global food security: Challenges and policies'. *Science,* 302(56542): 1917–1919. DOI: 10.1126/science.1092958.

Shafer, S.R., Walthall C.L., Franzluebbers A.J., Scholten M., Clark H., Meijs J., Reisinger A., Kazuyuki Y., Slattery B., Campbell I., Mcconkey B., Angers D., Soussana J. and Richard G. (2011) 'Emergence of the Global Research Alliance on Agricultural Greenhouse Gases', *Carbon Management,* 2(3): 209–214. DOI: 10.4155/CMT.11.26.

Schrader, G. and Unger J.G. (2003) 'Plant quarantine as a measure against invasive alien species: The framework of the International Plant Protection Convention and the plant health regulations in the European Union', *Biological Invasions,* 5(4): 357–364.

Scott, J. (2007) *The WTO Agreement on Sanitary and Phytosanitary Measures: A Commentary Oxford Commentaries on the GATT/WTO Agreements*, Oxford: Oxford University Press.

Swanson, T. (1997) *Global action for biodiversity: an international framework for implementing the Convention on Biological Diversity*, Earthscan: London, UK.

Watkins, L. (2013) 'NZ Pork fights flying pigs', *NZ Food Technology*, [Online], Available: http://foodtechnology.co.nz/content/nz-pork-fights-flying-pigs [10 May 2014].

Westmore, P. (2010) 'Quarantine: WTO rules in favour of NZ apples', *News Weekly*, 4 September, [Online], Available: http://newsweekly.com.au/article.php?id=4578.

Wilmshurst, J.M., Anderson J.A., Higham T.F.G. and Worthy T.H. (2008) 'Dating the late prehistoric dispersal of Polynesians to New Zealand using the commensal Pacific rate', *Proceedings of the National Academy of Sciences if the United States of America,* 105(22): 7676–7680. DOI: 10.1073/pnas.0801507105.

Wilson, J. and Riley S. (2012) '*State of the Tourism Sector 2012*', *State of the Tourism Sector No. 2*, Tourism Industry Association New Zealand (TIA) and Lincoln University. http://hdl.handle.net/10182/4554.

Worthy, T.H. and Holdaway R.N. (2002) *The Lost World of the Moa: Prehistoric Life of New Zealand.* Bloomington, IN: Indiana University Press.

CHAPTER 13

Japan's Science and Technology Diplomacy

Atsushi Sunami, Tomoko Hamachi, and Shigeru Kitaba[1]

Is Japan back? Since Prime Minister Abe took office in December 2012, he has emphasized the need to rebuild Japan's diplomacy in the face of unsolved critical issues in several important areas that affect relationships with the United States, China, and South Korea, among others. Prime Minister Abe wishes to regain Japan's confidence in the world as a leading country — 'like it used to be in the 1980s' — through both his economic policy ('Abenomics') and diplomacy strengthening alliances with the countries with 'common values'.

Since the beginning of Japan's two decades of stagnating economic growth in the 1990s, Japan's science and technology (S&T) infrastructure has faced many challenges. The nation's population is declining, which will likely reduce economic growth even further and therefore probably decrease both the amount of investment in S&T and the number of researchers working in the field. Additionally, the emergence of Asian countries in S&T, like Korea, China and India, has been remarkable over the last few years. Consequently, it is highly likely that Japan's relative strength in technology will continue to erode in today's globalized world.

Against this backdrop, there has been an increased interest in science and technology diplomacy in Japan. For Japanese policy makers, one of the primary objectives of S&T diplomacy is to tap into the growing science base beyond its nation's borders, including research facilities and human resources. International mobility of human resources for science is sometimes referred to as the 'brain circulation' that drives today's global science. Japan cannot

[1]This chapter has been rewritten based on the earlier version of their article, 'The Rise of Science and Technology Diplomacy in Japan', *Science and Diplomacy*, 2(1): 48–61.

allow itself to fall outside of the research network created by this brain circulation. Thus, an important objective for Japan's S&T diplomacy is to remain one of the critical points in this global network. However, Japan is opening itself up to the rest of the world rather too slowly in comparison to its neighbouring nations, such as Korea and China. By utilizing science and technology diplomacy, Japan can expand its volume of international research collaborations with dynamic nations around the world and can revitalize its innovation system by tapping into expanding global resource base.

The Abe Government and Science and Technology Diplomacy

Prime Minister Abe has been rather cautious in conducting his foreign policy so far. He has spent much of his time fixing the Japanese economy with his 'three arrows'. In his first year, Prime Minister Abe seems to have made reasserting Japan's industrial competitiveness through innovation his priority. His growth strategy contains three themes: (1) Reconstruction of Industry; (2) Creation of Strategic Markets; and (3) Global Strategy. For the reconstruction of industry, the emphasis is on the exit and entry of firms, which encourages start-ups through innovation and S&T. For the creation of strategic markets, the emphasis is on the role of innovation in the areas of life sciences and new energy. Finally, for the global strategy, the government will aggressively engage Japan in the dynamics of the global economy through the discussions surrounding the Trans-Pacific Partnership (TPP) and Economic Partnership Agreements (EPAs), using the external pressure, *gaiatsu*, in some sense to change the entrenched domestic institutions inhibiting Japan's globalisation efforts.

Japan must open up its innovation system to the rest of the world. In particular, the global S&T community is watching whether Japan will finally become ready to cooperate and ready to fully participate in the global innovation system. For example, while there have been discussions regarding what Japan should do to encourage more foreign students into Japanese research institutions under the Abe Cabinet, the number of foreign students is far from increasing, and Japan remains behind compared with China and Korea (see Figure 1). There are structural barriers in Japanese society derived from culture that make it hard for foreign researchers to adjust to our society outside of the laboratory (Van Noorden, 2012), but more foreign students and

No. of Foreign Students, China, Korea, Japan

(単位:人)

	JAPAN	China	Korea
India	541	10,237	548
Russia	333	14,971	307
Thailand	2,167	16,675	245
Africa	1,106	27,052	979
TOTAL	137,756	328,330	60,587

MEXT作成

Figure 1. Number of foreign students in China, Korea and Japan.
Source: Ministry of Education, Culture, Sports, Science and Technology, Japan.

researchers within Japan would help to increase the overall standard of Japanese research institutions, including universities.

In the face of mounting frustration among the science policy community, the Abe administration has embarked upon reforms to meet such challenges head on. He began working to re-establish the Council for Science and Technology Policy (CSTP), the headquarters for S&T policy in his cabinet office. As the result, Cabinet Office Establishment Act was partially amended so that CSTP can exercise more leadership over establishing innovation-friendly nation, and CSTP was renamed Council for Science, Technology and Innovation (CSTI) in May, 2014. By coordinating the ministries involved, the headquarters can carry out more aggressive S&T diplomacy to link Japan's S&T institutions with those from outside. Furthermore, Foreign Minister Kishida has established a special study group on Science Diplomacy in the Ministry of Foreign Affairs (MOFA) for the first time on the subject. As a result of the study group, Minister Kishida may create our first Science Advisor in the Ministry.

Japan's History with Science and Technology Diplomacy

On 19 August 2011, the Japanese government issued the fourth Science and Technology Basic Plan, a five-year national strategy on science, technology, and innovation outlining the goals for the coming decade. This is a notable step, as it was the first basic plan that designated S&T diplomacy as an issue of national importance.

Japanese Science and Technology Diplomacy before 2008

It was at the 66th session of CSTP in April 2007, chaired by then Prime Minister Shinzo Abe, that executive members of the Cabinet Office's CSTP issued a proposal titled, 'Toward the Reinforcement of Science and Technology Diplomacy'. The Japanese government has been formulating S&T basic plans since 1996, revising them every five years, and the CSTP members summarized the proposal during the period of the 3rd S&T Basic Plan in the hope that the nation would become aware of the increasing importance of collaboration between S&T and diplomacy, and Japan would increase its presence in the world.

As the word 'reinforce' suggests, even before the CSTP issued the 2007 proposal the notion of utilizing S&T to strengthen Japan's presence in the world had taken root. Under the first Basic Plan, effective from 1996 to 2000, the Japanese government hoped to thoroughly rejuvenate the nation's R&D system. As one of the measures to achieve the aim, it proposed the creation of a system designed to facilitate cooperation and exchange between the sectors, regions and countries included in the basic plan. In particular, the government listed a number of steps for Japan to take in order to enhance cooperation and exchange between nations:

1. Take a leading role in, and actively engage with, the promotion of international joint R&D;
2. Expand S&T cooperation, especially with developing countries; and
3. Build and create international R&D hubs in Japan which are attractive and open enough to gather excellent researchers from abroad as well as domestically.

With regard to the first goal, the government suggested that Japan 'provide a boost for existing projects such as the International Space Station Program, International Tokamak Experimental Reactor (ITER) Program, the Large Hadron Collider Program, and the Ocean Drilling Program (ODP)' and 'positively promote R&D in scientific fields on [the] global environment, ocean science [and] information/communications, where international cooperative systems are being created for projects, driven by the growing need for such cooperation' (Government of Japan, 1996). Regarding the second goal, the plan makes particular reference to the Asia-Pacific as an area of increasing importance in S&T cooperation. Even at this stage, Japan strongly intended to actively participate in global networks and establish cordial relationships with neighbouring countries.

In the second S&T Basic Plan (2001–2005) (Government of Japan, 2001), the government outlined its vision for Japan as an advanced science- and technology-oriented nation. This vision had three core components, aimed at making Japan:

1. A nation that contributes to the world by creating and using scientific knowledge;
2. An internationally competitive nation capable of sustainable development;
3. A safe, secure nation where people enjoy a high quality of life.

The plan declared the necessity of utilising S&T not only to solve difficult challenges that the world faces but also to secure Japan's international standing and national security, stating that Japan should take a leadership role in a system which contributes to economic development and the protection/standardisation of intellectual property rights, as well as in the advancement of technology itself. Asian countries were again referred to as the target for international cooperation, and also as the target for strategic cooperation in such standardization activities. Thus, while there are no direct references to diplomacy in those plans, it seems that the government was conscious of the necessity of building up an international cooperation framework in the field of S&T.

In accordance with the Basic Plans, Japanese universities and research and development (R&D) institutions conducted international joint research projects and academic exchanges with scientists from foreign institutions in various areas of S&T. In 2001, a program designed to allow Japanese R&D institutions to demonstrate international leadership was launched under the direction of CSTP. On a governmental level, Japan held policy dialogues with ministers and senior officials in charge of S&T; increasingly, as the

21st century dawned, with those from Asian countries. From the perspective of diplomacy, S&T contributed to building good relations with other countries. For example, Japan concluded 24 agreements on scientific and technological cooperation with 34 countries by 2000. (At present, there are 32 agreements with 46 countries and the European Commission.) The first was concluded in 1973 with the former Soviet Union, and agreements concluded in the 1970s include those with Central and Eastern Europe countries and newly independent states (i.e., parts of the former Soviet Union). It is no exaggeration to say that S&T, a borderless field for the pursuit of truth, plays an important role in promoting trust among nations.

Moreover, S&T has been an effective means for establishing a trusting relationship with developing countries. Since the mid-1950s, Japan has contributed to the improvement of social development and the welfare of people in developing countries through Official Development Assistance (ODA), and through its framework many Japanese S&T researchers have been dispatched to developing countries over the years.

Thus, international activities in the field of S&T, whether they originated from scientific or diplomatic interests, have steadily contributed to maintaining Japan's strong presence in both the S&T and diplomatic communities. However, people in the S&T sphere have, in the past, often given little consideration to diplomacy when they collaborated on international projects. Likewise, Japanese diplomats have not often thought about using Japan's S&T as a diplomatic tool. In other words, S&T and diplomacy have not been strategically linked to one other.

The Emergence of Science and Technology Diplomacy in Japan

The concept of science and technology diplomacy was first publically recognized in 2008 with the release of the CSTP report, 'Toward the Reinforcement of Science and Technology Diplomacy' (Council for Science and Technology Policy, 2008). The report, based on the discussions of a CSTP working group that operated from July 2007 to April 2008, defines science and technology diplomacy as any steps taken 'to link S&T with foreign policy so as to achieve their mutual development' and 'to utilize diplomacy for the further development of S&T and promote efforts to utilize S&T for diplomatic purpose'. It also describes a number of basic strategies for promoting science and technology diplomacy within Japan:

1. Establishing systems in which Japan and its counterparts can enjoy mutual benefits;
2. Generating synergy between S&T and diplomacy for resolving the global issues facing mankind;
3. Developing "human resources" that sustain S&T diplomacy; and
4. Increasing Japan's international presence.

The report insists that Japan's S&T diplomacy place importance on strengthening both S&T cooperation with developing countries in order to resolve global issues, and S&T cooperation with developed countries using Japan's advanced S&T capabilities.

Why did the concept of science and technology diplomacy come to public attention at that time? One factor behind its emergence was the necessity of demonstrating leadership in a series of important international gatherings that were held in 2008. Japan hosted the G8 summit, G8 related ministerial-level meetings (including the G8 S&T Ministers' Meeting), and the Tokyo International Conference on African Development IV (TICAD IV). In the midst of the accelerating growth of emerging economies, it was time for Japan to make the most of S&T as a source of 'soft power' (Nye, 2004).

However, this reinforcement of science and technology diplomacy was not only intended to meet diplomatic needs. The other trigger was the need to open up Japan's science community to the world and end the inward-looking propensity of Japanese researchers. The CSTP working group report was unique in that it encouraged S&T cooperation with developing countries as well as developed countries. Cooperating with developing countries is helpful to solving global issues, and at the same time the working group considered it valuable to the revitalization of Japan's science community. The working group thought that foreign researchers in dynamic nations could stimulate Japanese researchers and be a positive influence on Japan's R&D system, as it was in the era of 'catching up' during the last century.

Developing the Concept of Science and Technology Diplomacy

The idea of science and technology diplomacy gradually spread among the S&T and diplomatic communities in Japan. At the same time, however, as the 'catching up' nations rapidly increased their presence, it became difficult for Japan to maintain a leading position in the field of S&T. For example, China and South Korea's share of research papers in 22 S&T fields cited in Thomson Reuters' *Web of Science* has been increasing since around 2000, while that of Japan has been declining over the same period. The growth of Chinese and South Korean articles in the share of the top 10 percent of research papers in these 22 fields is also remarkable. According to a report by the National Institute of Science and Technology Policy in Japan, China has increased its presence as the USA, UK and Germany's primary partner in writing joint papers (Saka and Kuwahara, 2010). Furthermore, in recent years Japan has been facing the stagnation of R&D expenditures both in government and in the private sector. Taking these trends seriously, the executive members of the CSTP made a proposal to further strengthen Japan's science and technology diplomacy in June 2009 (Taizo, 2009).

Following this proposal, the CSTP created a task force to identify concrete measures to strengthen Japan's role in the world while considering how the world will change by 2020. The forecast shows that the erosion of Japan's relative strength in science is almost inevitable. For example, it is estimated that the percentage of the population aged between twenty and thirty-nine will reduce to almost three quarters of its 2005 value by 2020. This will likely also lead to a decline in the number of researchers and scientists who lead Japanese S&T. Another estimate shows that Japan's share of global R&D expenditure will be four points down by 2020 (compared to 2006).[2]

Taking this vision of the future into account, the task force compiled a tough-minded report in February 2010. Unlike the 2008 report, the 2010 report points out that some developing counties are no longer just the recipients of technology but are on an equal footing, and that therefore Japan should integrate its R&D system with R&D resources in the rest of the

[2]This estimation was calculated based on data in the United Nations World Population Prospects, the 2008 Revision (UNWPP, 2008).

world, including those in developing nations. Another point worth making is the shift in the perception of 'diplomacy'. The task force suggested that 'diplomacy' should not just involve establishing good relationships with other nations, but should also encompass the realization of national interests and the strengthening of the industry's international competitiveness. Thus, science and technology diplomacy came to be seen as a more strategic aspect in the revitalization of Japan.

The 4th Science and Technology Basic Plan (August 2011) (Government of Japan, 2011) embraces the key points made in the task force report. The plan recommends that Japan strategically develop its international S&T activities together with dynamic nations. Based on that idea, and aware of the energy of growing Asian nations, the plan proposes that the Japanese government promote the East Asia Science & Innovation Area initiative: an initiative designed to promote open regional cooperation in which countries collaborate in promoting cross-border flows of people, goods, and capital to enhance R&D efforts aimed at addressing common issues in Asia. The task force proposed the initiative in the expectation that science, technology, and innovation would help build a more integrated East Asian community proposed by then Prime Minister Yukio Hatoyama. The East Asia Science and Innovation Area initiative is new in that S&T is clearly positioned as a source of diplomatic soft power.

Now Japan's science and technology diplomacy enters a new phase, advancing from just transferring technologies or R&D results overseas to strategically using S&T for diplomacy and leveraging diplomacy to help strengthen Japan's S&T infrastructure.

Representative Measures Taken under the Science and Technology Diplomacy Initiative

Since formalizing the science and technology diplomacy concepts from 2008 to 2010, Japan has taken several important steps to strengthen S&T cooperation with dynamic nations around the world. These measures are inter-related in an attempt to promote: (1) joint research with developing countries in order to resolve global issues as well as provide capacity building in those countries; (2) research cooperation in the field of cutting-edge technology with technologically advanced countries; and (3) cooperation based on an equal partnership with East Asian countries in the context of the East Asia Science and Innovation Area.

Research Cooperation with the Developing Countries

In accordance with the 2008 CSTP report, which emphasizes S&T cooperation with developing countries, the Ministry of Education, Culture, Sports, Science and Technology (MEXT) and the MOFA launched a new program entitled Science and Technology Cooperation on Global Issues in 2008. This program has two subprograms: the Dispatch of Science and Technology Researchers program and the Science and Technology Research Partnership for Sustainable Development (SATREPS) program.

The Dispatch of S&T Researchers program has been jointly operated by the Japan International Cooperation Agency (JICA) under the umbrella of MOFA, and the Japan Society for the Promotion of Science (JSPS) under the supervision of MEXT. In this program, according to the needs of the partner countries, the most suitable researchers in Japan are dispatched to developing countries as JICA experts to engage in joint research that MEXT and JSPS select. This program aims to make significant international contributions through joint research that is expected to develop new technologies and to enhance the research capacity of Japan and its counterpart countries.

SATREPS is another promising program for promoting joint research with developing countries. In SATREPS, the Japan Science and Technology Agency (JST) and JICA collaborate to promote international joint research that targets global issues — such as limited bio-resources, natural disaster prevention, and infectious disease control — that are based on the needs of developing countries. It also aims to promote international joint research that includes a plan for future social implementation by collaborating with ODA in order to acquire new knowledge that will lead to solutions to global issues and advance the level of scientific and technological capacity in developing countries.

SATREPS projects are selected each year from project proposals submitted by Japanese research institutions. JST uses research contracts to support research costs incurred in Japan. JICA provides support through its technical cooperation project framework to cover costs in the developing country. The overall R&D management of the international joint research is handled jointly by JST and JICA. JST has the expertise in funding research projects at research institutions in Japan, and JICA brings experience in technical cooperation in developing countries (Figure 2). Since it began in April 2008, a total of 78 SATREPS projects have commenced in 39 countries.

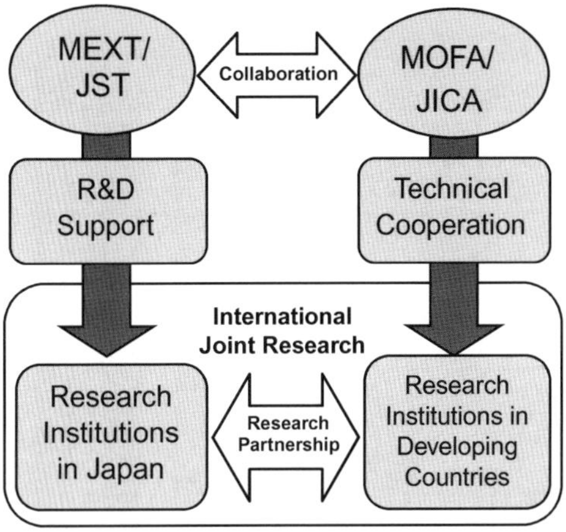

Figure 2. Scheme of implementation of 'SATREPS'.

This type of collaboration between funding agencies and foreign development agencies has also been seen in the United States where the US Agency for International Development and the National Science Foundation launched Partnerships for Enhanced Engagement in Research (PEER).

Thus, Japan has been increasing its willingness to open up its scientific programs to foreign partners and to sponsor genuinely collaborative partnerships with developing countries.

Research Cooperation in the Field of Cutting-Edge Technology with Technologically Advanced Countries

The Japanese government is aware that for Japan to achieve a world-class S&T capability in such an intense economic and technologically competitive environment, it is more important than ever for the government to manage international joint research in a strategic manner. This means MEXT must designate countries and research fields of cooperation in a top-down manner, on the basis of intergovernmental agreements.

In line with this government policy, JST has been implementing a research exchange program known as the Strategic International Cooperative Program (SICP) since 2003. This program provides intensive support to mostly advanced countries with relatively small international research projects.

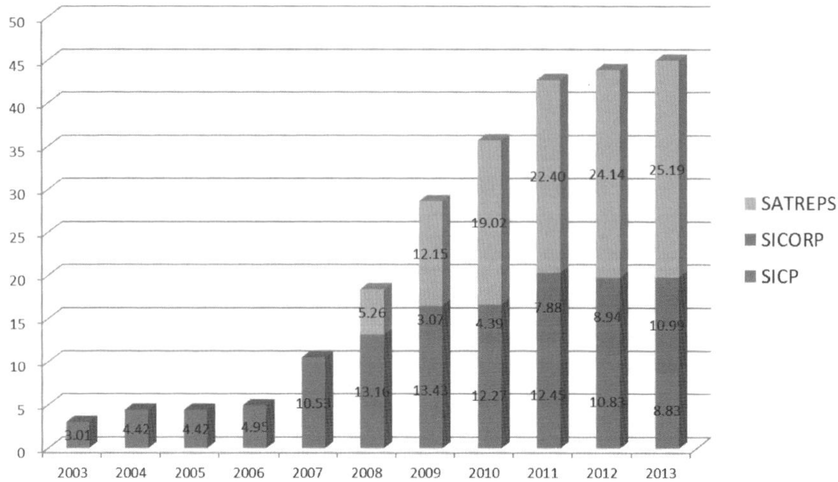

Figure 3. Budget transition of JST's international research cooperation programs (Million US$, 1US$ = 95 JP¥).

Source: Department of International Affairs, Japan Science and Technology Agency.

Aiming to further develop science and technology, JST has supported 383 research projects in twenty-two countries and one region (as of June 2013).

In addition to SICP, JST also started a new program for funding relatively large international joint research projects. Since 2009 this Strategic International Collaborative Research Program (SICORP) has seen its budget increase substantially (Figure 3) and has supported fourteen ongoing projects in five countries and one region. The CSTP task force report in 2010 — which focused on integrating domestic R&D resources with those in technologically dynamic nations in order to maintain the relative strength of Japan's S&T capacity — supported SICORP, which intends to promote international research cooperation with technologically advanced countries.

Cooperation Based on Equal Partnerships with Asian Countries

As has already been mentioned, one of the policy goals in Japan's science and technology diplomacy is to strengthen the domestic R&D system by integrating foreign R&D resources. The Japanese government's initiative to build an East Asia Science & Innovation Area, which is aimed at raising the capability of R&D and addressing common problems in the region, is one of those attempts, and the East Asia Joint Research Program (e-ASIA JRP) represents the main part of the initiative.

The e-ASIA JRP is Japan's proposal for developing and supporting joint research projects in East Asia on a multilateral basis. Prospective members of the program include public funding agencies and governmental bodies of countries participating in the annual East Asia Summit (EAS). The multinational research collaboration is designed to be managed by a 'matching fund system', in which support from each ministry or agency will go to national universities or research institutes in each country.

This multinational research collaboration program is multipurpose. The promotion of multilateral joint research in fields such as life sciences, green technology, and disaster prevention is intended to contribute to the resolution of shared regional challenges. The improvement of scientific and technological capabilities is expected to have a positive effect on the further development of the region, which is at the centre of global economic growth. From a diplomatic point of view, Japan can expect to play an active role in strengthening mutual trust and benefits among countries in the region.

At the sixth EAS meeting held in November 2011 in Bali, Indonesia, the chair's summary stated: 'We welcome Japan's initiative for implementing the e-ASIA JRP/multilateral joint research program under the concept of East Asia Science and Innovation Area'. The e-ASIA JRP was formally inaugurated at the first board meeting held in Singapore in June 2012. The founding members included the S&T related ministries from eight countries: Indonesia, Japan, Laos, Malaysia, Myanmar, the Philippines, Thailand, and Vietnam. As of May 2014, 13 institutions in 11 countries including the United States, Cambodia and New Zealand, participate the e-ASIA JRP.

However, the initiative has just begun and there remain several challenges that need to be addressed. The most obvious one is that it needs to include more countries with ties to the region, especially India, South Korea, Singapore, and China. Without the involvement of the most dynamic countries with robust R&D resources, the e-ASIA initiative will not be able to live up to its potential. Also, it may need to harmonize existing programs, such as the ASEAN Committee on Science and Technology (COST) and the Asia Pacific Economic Cooperation (APEC) forum's Industrial Science and

Technology Working Group. Whether this S&T initiative will succeed or not depends on Japan's capability to coordinate national interests among countries; that is, diplomacy.

Challenges and Opportunities

The Japanese government has developed a program that directly links to its diplomatic strategy and could lead to the integration of Japan's R&D system with other countries in the growing science base. However, Japan's science and technology diplomacy still has some issues. There is still a lack of connectedness between S&T policy and foreign policy. For example, Minister of Foreign Affairs has not yet been included in CSTI member. Concerning process of planning of foreign policy making, there are not many staff with S&T related backgrounds. Even though many programs like SATREPS and e-ASIA JRP help to bridge that disconnect, they have not been fully exploited as solutions to diplomatic issues, such as economic diplomacy or resource security. Likewise, people on the S&T side fall short in using diplomacy to strengthen Japan's research and development system. Japan's science and technology diplomacy produced a synergistic effect in both sides yet.

Another challenge is that most Japanese political leaders do not perceive S&T as a useful instrument for foreign policy. Even if they do, they rarely mention it in international fora. However, this situation has been gradually changed under the second Abe Cabinet. Three months after the establishment of Abe Cabinet in December 2012, the prime minister ordered formulation of 'Comprehensive Strategy on Science, Technology and Innovation' at the 107th session of CSTP. Then, on 7 June 2013, the CSTP-compiled report 'Comprehensive Strategy on Science, Technology and Innovation — A Challenge for Creating Japan in a New Dimension' (Cabinet Office, 2013: 2, 6) was approved by the Cabinet. The report reflected the need to '[come] out with the whole picture of the science, technology and innovation policies as problem-solving strategy packages' and has three central characteristics:

1. The Strategy consists of a long-term vision and immediate action program;
2. The Strategy is a comprehensive package of problem-solving science, technology and innovation policies;

3. The Strategy is formulated for the nation as a whole, including researchers, companies, universities, research institutions, and the general public.

Promotion of science, technology and innovation (STI) was further positioned as a pillar of the Industry Revitalization Plan in 'Japan Revitalization Strategy — JAPAN is BACK', approved by the Cabinet on 14 June. Although it will take some time before Japan uses science to address diplomatic issues, STI is increasingly included in the statements of top- or high-level meetings, such as at the Visegrad Group (V4) plus Japan Summit Meeting in June 2013 and TICAD V, as a prioritized area of cooperation.

It is time for Japan to reaffirm its global significance. To save Japan's science from its relative decline in this rapidly changing world, the Abe administration should think hard about how to more firmly incorporate science and technology into Japan's foreign policy. It is time for Japan to seriously consider science diplomacy as an important tool for expanding the frontier and adding breadth to our diplomacy. When Japan's foreign relations face an unprecedentedly difficult period, science diplomacy may find its place inevitably in the centre stage of Japan's foreign policy.

References

Cabinet Office (2013) 'Comprehensive Strategy on Science, Technology and Innovation — A Challenge for Creating Japan in a New Dimension', *Government of Japan*, 7 June, [Online], Available: http://www8.cao.go.jp/cstp/english/doc/20130607cao_sti_strategy_provisional.pdf [14 May 2014].

Council for Science and Technology Policy (2008) *Toward the Reinforcement of Science and Technology Diplomacy*, 19 May, [Online], Available: http://www8.cao.go.jp/cstp/english/doc/s_and_t_diplomacy/20080519_tow_the_reinforcement_of.pdf [14 May 2014].

Government of Japan (1996) *Science and Technology Basic Plan*, 2 July 1996, [Online], Available: http://www8.cao.go.jp/cstp/english/basic/1st-BasicPlan_96-00.pdf [14 May 2014].

Government of Japan (2001) *The Science and Technology Basic Plan (2001–2005)*, 30 March, [Online], Available: http://www8.cao.go.jp/cstp/english/basic/2nd-BasicPlan_01-05.pdf [14 May 2014].

Government of Japan (2011) *Science and Technology Basic Plan*, 19 August 2011, [Online], Available (in Japanese): http://www8.cao.go.jp/cstp/kihonkeikaku/4honbun.pdf [14 May 2014].

Nye, J. (2004) *Soft Power: The Means to Success in World Politics*, New York: Public Affairs.

Saka, A. and Kuwahara T. (2010) *Benchmarking Scientific Research 2010 — Biblio-metric Analysis on Dynamic Alteration of Research Activity in the world and Japan*, 15–17 December, NISTEP, MEXT, Japan, Available (in Japanese): http://data.nistep.go.jp/dspace/bitstream/11035/909/7/NISTEP-RM192-FullJ.pdf.

United Nations World Population Prospects (UNWPP) (2008) *United Nations World Population Pro-spects: The 2008 Revision*, [Online], Available: http://kczx.shupl.edu.cn/download/786444c9-20c1-4b5a-b0d6-d7544569a2ee.pdf

Van Noorden, R. (2012) 'Global Mobility: Science on the move', *Nature*, 490: 326–329. DOI:10.1038/490326a.

Yakushiji, Taizo (2009) 'The Potential of Science and Technology Diplomacy', *Asia-Pacific Review*, 16(1):1–7. DOI: 10.1080/13439000902957640

Conclusion

Chapter 14

New Day or False Dawn?

Lloyd S. Davis and Robert G. Patman

The Modern Era of Science Diplomacy

The preceding chapters demonstrate unequivocally that 'science' has indeed become entwined with 'diplomacy' to such an extent that it represents, justifiably, its own designation as a recognisable and legitimate form of diplomacy. If this were literature, it would occupy its own genre; if this were biology, it would be its own species.

The abiding question, then, is not does science diplomacy exist? Or even, given that, what form does it take? Those questions have been elaborated upon fully and frankly in the contributed chapters to this volume. The issue of concern, which we are left with, is the one suggested by the title of this book: how significant is science diplomacy in the context of international relations? Is it, indeed, a new day or a false dawn that we see on the horizon of global politics? And what does this mean for a growing array of international problems that require scientific input?

To the extent that science has been linked, in one way or another, to diplomacy for centuries, if not more, it may not be considered a new day at all. However, as noted by Turekian *et al.* (in Chapter 1), it is the strength of the involvement of science in diplomacy over the last two-and-a-half decades since the end of the Cold War, coupled with the rapid expansion of globalisation brought about by revolutionary developments in information and communication technologies, that seems to be qualitatively changing the relationship between science and diplomacy.

The real driver for this — in fact, it is the very heart of the matter — is that the major problems and issues facing sovereign states are of global proportions and almost every one of them is, in some way or other, connected to science and technology: climate change, water conservation and pollution, pandemics, the exploration of space, the loss of biodiversity, food production, trade, economic performance, diminishing energy resources, our understanding of quantum physics, drones and wars fought with high-tech weaponry, how to provide data security, and how to protect privacy. None of these challenges can be resolved adequately or meaningfully by a single country. But there is a paradox here. While the number of national problems requiring international scientific solutions is rapidly growing, many sovereign states remain in denial about this and so the international means for addressing these challenges remains weak and incomplete (Held, 2010).

The modern era of science diplomacy, therefore, has emerged during what is likely to be a lengthy international transition. Diplomats need to be guided by science to deal with the pressing issues of the day (Science for Diplomacy); the way for science often needs to be leavened by diplomats (Diplomacy for Science); and sometimes diplomats can use science for other ends (Science in Diplomacy). That there is a need for science diplomacy in today's world is undeniable, but there remain serious doubts as to how effective it can be — whether it really can be a panacea for some of the difficulties that countries face — and a clear reluctance by states to accept that the old doctrine of unfettered sovereignty should be modified to facilitate more effective science diplomacy in an increasingly interdependent world.

The Limitations of Science Diplomacy

Climate change, as a consequence of anthropogenically induced global warming, is the poster child for issues that transcend the boundaries of states and can only be tackled meaningfully by an international effort. In 1988, there was the formation of the Intergovernmental Panel on Climate Change (IPCC), an international group of more than 2,000 scientists that coordinated and reviewed the available evidence and data in the field. They concluded that not only was global warming a significant threat — perhaps *the* most significant threat — to the future of the planet, they also identified the culprit too: increased CO_2 in the atmosphere as a consequence of activities undertaken by humans. These included emissions from industrial pollution, traffic emissions, and intensive farming of ruminants.

To be sure, diplomatic efforts led to international treaties designed to tackle climate change and limit greenhouse gas emissions at several high profile meetings of the United Nations Framework Convention on Climate Change (UNFCCC): Rio, Kyoto, Bali, Copenhagen, Cancún, Durban and Doha. In the build-up to these meetings, scientists or their representatives spelt out consequences of inaction with respect to climate change. However, as Manjana Milkoreit makes clear in her contribution (Chapter 6), many of those charged with the task of negotiating — the diplomats — did not have a full grasp of all of the scientific concepts underpinning what was at stake, nor an ability to conceptualize what the long-term scenarios predicted by the scientists' data and models really meant. Mitigation protocols were proposed and debated but eventually, and perhaps predictably in the circumstances, much of the substance of the discussions was watered down in order to maintain some form of diplomatic consensus.

The 1997 Kyoto Protocol set emissions targets for developed countries that were supposedly binding under international law although, notably, the USA (the world's second-largest emitter) did not ratify the protocol. In the years after Kyoto, atmospheric CO_2 levels continued to rise, surpassing 400 parts per million (ppm) in recent times[1]. The UNFCCC Meeting in Copenhagen in December 2009 was widely seen as the world community's urgently needed opportunity to renegotiate the Kyoto Protocol and find a consolidated approach to tackling the issue of greenhouse gas emissions.

In the end, little was achieved (Backstrand and Elgstrom, 2013; Oels, 2013). A so-called Copenhagen Accord was developed by some of the countries, which called for emission reductions sufficient to limit the overall mean global temperature rise to less than 2°C relative to pre-industrial levels (Luderer *et al.*, 2013). Representatives and diplomats went back to their respective states charged with enacting the necessary legislation so that individual countries could meet their obligations and targets for emission reductions, albeit with no measures of ensuring compliance beyond the good intentions of the nations involved (Glomsrod *et al.*, 2013).

The targets set by the countries through this consensus oriented diplomatic process were woefully inadequate according to the scientists' predictions

---

[1]To see an up-to-date weekly average CO_2 reading, visit http://www.esrl.noaa.gov/gmd/ccgg/trends/weekly.html.

and models of the levels of change that must be enacted to avert disaster from climate change (Buhr *et al.*, 2014; Rogelj *et al.*, 2013; Luderer *et al.*, 2013). Even then, most countries have struggled to even get close to their targets (Hof *et al.*, 2013; Hope and Hope, 2013). The situation has been described as 'international anarchy' and some argue that the only way forward is a system that allows countries like China to pollute much more than other states such as the US (Shum, 2014), even though China in recent years surpassed the US as world's largest emitter, accounting for over a quarter of all the world's CO_2 emissions (data from the US Carbon Dioxide Information Analysis Centre[2]) — a proportion that is inevitably going to increase (Jiang *et al.*, 2013). Climate change politics have become dominated by antagonisms and disagreements, especially between the developing countries and the developed countries, over the question of the historic responsibility for the problem of global warming (Chatterton *et al.*, 2013; den Elzen *et al.*, 2013; Hallding *et al.*, 2013).

In the case of climate change, then, it would seem that science diplomacy has so far not proved up to the task. More a case of a false dawn than a new day. The perceived self-interest of individual countries has overridden the need for an outcome that can only be generated through the recognition of an urgent collective responsibility. It is Garrett Hardin's *Tragedy of the Commons* played out on the largest stage of all: the globe (Hardin, 1968).

Science, truth and facts were not enough to trump the traditional drivers of international relations: the Westphalian doctrine of absolute state sovereignty and the corresponding belief that there is no higher authority than the state for defining national economic and diplomatic interests. This provides a clue as to one of the main challenges for science diplomacy: it is not enough for scientists just to inform diplomats (or the representatives of a state) of the relevant scientific facts. Instead, those scientific facts and information need to be translated into language that the diplomats and the country's representatives can readily understand in a national and international context. In essence, the problem faced by politicians and diplomats is the same one faced by the public when it comes to science: they find it difficult to understand.

But, according to Joan Leach (Chapter 8), this is where science communication can help. For diplomats to be able to speak confidently and

---

[2] For the latest estimates, see http://cdiac.ornl.gov.

accurately about science, the language of science needs to be made less dense and impenetrable. It needs to be expunged of jargon. Concepts need to be simplified while maintaining their veracity. The degree of certainty needs to be emphasised rather than the degree of vagueness. The scientists' world of probability values needs to be expressed in ways that do not imply vagueness and uncertainties, because these will always be used by governments as excuses for inaction.

Aside from climate change, there have been other areas where the potential to do good from marrying science to diplomacy has also been an abject failure. Sefton Darby (Chapter 7) argues compellingly that diplomacy has hindered rather than helped the science and tech-heavy oil and mining industries. The fiasco over the financing and operation of the oil pipelines from Chad and Azerbaijan exemplify how the realities of economics and corruption in governments can lead to unsatisfactory outcomes (outcomes that, ironically, exacerbate the difficulties of many countries seemingly blessed with oil and mineral resources — the so-called *resource curse*), especially when the international community, for political expediency, turns a blind eye to such corruption. In a sense, these examples underscore that science diplomacy cannot be a replacement for other forms of diplomacy. For example, it is not a substitute for diplomacy that seeks to resolve essentially political issues such as national self-determination, border disputes, or justice for ethnic minorities. Nevertheless, science diplomacy may be a useful complement to diplomacy designed to bridge political differences.

If, in some situations, the political aspect of diplomacy can complicate scientific and technological co-operation between states, it is also true that science and technology can be a problem for diplomatic co-operation. Daryl Copeland (Chapter 9) asserts that the very technology upon which nations rely for covert operations and spying can also be their undoing. In the Wikileaks case outlined by Copeland, US Army Private Bradley Manning was able to access and copy vast amounts of sensitive data and spirit them away to Wikileaks, the online non-profit journalistic organisation foundered by Julian Assange. Indeed, Wikileaks was set up primarily to expose much of the covert behaviour of governments that passes for diplomacy. The subsequent removal in 2013 of thousands documents from the US National Security Agency (NSA) by 29 year-old Edward Snowden, and their staged release to the likes of *The Guardian* newspaper in the UK

(Greenwald *et al.*, 2013), has exposed the degree to which the United States uses technology to covertly spy on other nations and even its own citizens. This has had major ramifications for diplomacy.

On the one hand, it has eroded the tacit social contract that existed between citizens and their government: that law-abiding American citizens should be free from surveillance by their own elected government. It has also by extension broken the trust between many of the United States' allies: Snowden's documents revealed that the United States is not just monitoring communications of hostile nations, but also those of their democratic allies and at the highest of levels. The Snowden revelations have raised the issues of privacy and security of personal information in the consciousness of citizens of many countries. This has fuelled both a focus on international relations and an attempt — at least in the short-term — from governments like those in the United States and the United Kingdom to become increasingly transparent about what data they collect, how they collect that data, how they store the data, and what they do with the data. There has been pressure on governments to enact legislation to limit the extent that individual citizens can have their privacy violated. For example, it is evident that President Obama plans to ask Congress to end the bulk collection and storage of phone records by the National Security Agency of calls made in the United States, allowing the government to access the 'metadata' from the phone companies only after getting permission from the Foreign Intelligence Surveillance Court (Rampton, 2014).

Thus, rapid developments in surveillance technology have the potential to undermine the political and diplomatic legitimacy of a government. At the same time, the examples of climate change and the experience of the oil and mining industries in certain countries serve as reminders that politically motivated diplomacy can get in the way of the science, and that other factors shape policy: national security concerns, desired access to resources, apparent economic self-interest, wars, and so forth.

In other words, there are often situations that fall under the banner of Science for Diplomacy in which the notion of science diplomacy has not produced the desired outcomes. On the other hand, science diplomacy has had some notable successes, most prominently in the area of Diplomacy for Science.

The Promise of Science Diplomacy

Where diplomacy has been used to help initiate and manage large-scale international science projects, it has often been a success. Examples that come to mind are the International Space Station, the Large Hadron Collider and, as detailed by Maria Pozza (Chapter 5), the Square Kilometre Array. While it is possible to think of these as simply the traditional science collaborations common to science, the scale of these projects and their ramifications require involvement by governments to work effectively. Not only does diplomacy facilitate science — science that is occurring on such a scale that it could not be enacted without the involvement and agreement of countries — but both the process of collaborating and the scientific outcomes can lead to benefits such as improved inter-state relations and greater international co-operation.

Antarctica is the one area of the globe that arguably is the archetype for what can be accomplished by Diplomacy for Science. Antarctica is more-or-less a large zone of peace set aside for countries to collaborate for largely scientific purposes. Since the International Geophysical Year (IGY) in 1957–1958, and the signing of the Antarctic Treaty in 1959, nations with a presence in Antarctica have largely been involved in a co-operative scientific endeavour. As Gary Wilson (Chapter 4) notes, this international co-operation is often driven from the ground up, with the scientists pushing the diplomats to support them. However, the reality is that from the beginnings of exploration of the Antarctic continent to the modern-day operations of national bases and the large-scale international research projects like ANDRILL, these could not have occurred without the support of governments, and often governments acting together. For example, the entire New Zealand Antarctic Programme, which is expressly set up to foster science, could not exist without the logistical support provided by the United States. Bilateral agreements, like those whereby landing rights in Christchurch for the American aircraft en route to Antarctica are exchanged in return for providing seats for New Zealanders, do not occur without the involvement of professional diplomats.

Antarctica should be rightly viewed as a major success for science diplomacy. But it equally shows the limits of international diplomacy to consistently deliver beneficial outcomes. The fishing of toothfish (*Dissostichus* spp.) in the Ross Sea region of Antarctica is a prime example: the fishing is being carried out at best with little knowledge of its effects on the ecosystem of the world's most pristine ocean, and at worst with devastating effects on a fish population that breeds very slowly (Ainley *et al.*, 2012). Exercising caution and banning or severely restricting fishing is the only policy that makes any rational sense. Yet, the international body set up to monitor and control the marine assets around Antarctica, the Convention for the Conservation of Antarctic Marine Living Resources (CCAMLR), consists of the representatives from 25 nations and relies upon a consensus-based mode of decision-making. As a consequence, it is open to being stymied by the opposition of just one renegade nation: in the case of a ban on fishing for toothfish in the Ross Sea, it has been Russia (Neslen, 2013).

Hence, once again, despite science diplomacy making a positive difference in the Antarctic, the continuing adherence by nations to an absolute doctrine of state sovereignty in international bodies like CCAMLR does not sit well with a commitment to the 'greater good' in an interdependent world and means that science diplomacy quickly runs up against limits in terms of what it can achieve.

In instances where science is used in the arsenal of diplomats to exert change and influence over other nations, it has had some limited success. Cathy Campbell (Chapter 2) demonstrated how Arab nations are by almost all measures doing poorly when it comes to being scientifically literate, conducting science, and publishing it. The United States has tried to exert some influence over such countries by using science projects as a tool for encouraging closer co-operation (Koenig, 2009). While in principle this means of diplomacy would seem to offer much in terms of being mutually beneficial, its reach and import are dependent upon being targeted (so as not to dilute its effect) and being adequately resourced. To use science effectively in diplomacy requires, then, first the identification of a specific goal, and second, adequate funding and resourcing to allow that goal to be achieved.

One area where science successfully acts as a means of improving diplomatic relations between countries, be they Arab nations (Sarkadi and Schatten, 2012) or other states, is in the area of health. As Edison Liu (Chapter 11) has shown in the case of Asian countries, a willingness to nurture the involvement of less science-endowed countries like Indonesia, Thailand and the Philippines into the fold of the Pan-Asian Genome Project, paid dividends for all participants by increasing the scientific prowess of those countries with the least developed infrastructure for this type of research, and also increasing the study's access to samples from populations with the most ethnic diversity. This helped the study achieve its overall goals, such as understanding the process of migrations throughout Asia — a phenomenon that goes well beyond the boundaries of single states.

In such examples, there is little potential downside for any participant in the diplomatic exchange. That is, it is a win-win situation for all concerned. However, that is not always the case when it comes to science and diplomacy. The relationship between biosecurity and trade is one issue about which agricultural nations, especially, are sensitive. On the one hand, such nations are dependent upon their exports to create wealth, but on the other, their means of sustaining their wealth generating industries could be put at risk by introduced pests or diseases. As noted by Goldson and Gluckman (Chapter 12), access to markets and trade interests need to be balanced against biosecurity risks. Again, is so often the case in the international arena, perceived self-interest can trump science when it comes to international politics and relations. For example, successive Australian governments since 1919 had invoked 'scientific reasons' for banning the import of New Zealand apples into Australia until the World Trade Organisation rejected the idea that the ban was based on scientific grounds and overturned the ban in 2010. The apple example underscores the value of having the science — and the risks — evaluated by a non-partisan international disputes resolution panel whose findings were binding on all parties. Without the WTO's ruling, it is highly likely the Australia's stance on banning New Zealand apples would have continued. Clearly, the assessment of scientific risk by states can be shaped by political considerations rather than by objective concerns.

Sarah Macindoe (Chapter 3) takes this point further in her discussion about managing the plant genetic resources used for food and agriculture. The perceived self-interests of states and businesses are juxtaposed with what she calls the *global commons*. Here it is science and technology that are being leveraged to categorise and protect the genetic material that forms basis for agriculture and feeding people worldwide. Countries like New Zealand, with expertise in modern genomic techniques and extensive botanical databases, are leading the way in helping other countries manage and conserve plant genetic resources. Hers is an optimistic view of the world perhaps, yet one that is similarly being promoted as a means for tackling problems associated with global warming (Chatterton *et al.*, 2013) — where diplomacy and science exist to make the world a better place for all — and Macindoe is the first to admit that the continued centrality of sovereignty and national interest to the international system means that a global perspective, in terms of what is best, will not always prevail.

As related, perceived self-interest is often what motivates what is classified as 'Science in Diplomacy'. Atsushi Sunami *et al.* (Chapter 13) contend that Japan uses its prowess in science and technology to help establish trusting relationships with developing countries: a laudable endeavour in itself that may even be considered 'altruistic' at first glance. However, Sunami *et al.* assert that what is driving this use of science in diplomacy is Japan's recognition that, since Tokyo's economic stagnation during the 1990s, the country is losing its standing as one of the world's premier science and technology leaders. In the areas of science, technology and innovation, Japan is facing stiff competition from other Asian countries, particularly China, South Korea, and India.

One motivator for the push to use science in diplomacy is so Japan can tap into science bases beyond its borders, thereby making sure that Japan remains a stop on the route of the 'brain circulation' cycle. The latter is a consequence of increased mobility of scientists, inventors and researchers and the ensuing competition for their services. International collaborations in science do not just accomplish things that would be beyond the knowledge and resources of a single country, they also expose and educate a country's scientists to ideas, equipment and the latest techniques in the rapidly developing areas of science and technology.

The Role of Scientists, UN Reform, and the Social Media in Diplomacy

While science for diplomacy has had a mixed record so far — one characterised by some disappointing results and measured successes — the picture is brighter with respect to large scale international diplomatic co-operation for science. Here, there have been some highly visible successes. Examples include the International Space Station, Antarctica, the Square Kilometre Array, and the Large Hadron Collider. When science co-operation is the desired outcome, diplomacy has often helped to facilitate successful outcomes. But when diplomatic co-operation is the desired outcome, science has often been less successful in generating positive outcomes. The continuing diplomatic failure to limit emissions causing global warming, strengthen biosecurity internationally, and diminish the impact of the so called *resource curse* in the oil and mineral sector in a number of countries, all point to this trend.

Yet the fact remains that the big issues facing the world stem, for the most part, from science and technology — and while feasible technical solutions may well be available for many of the problems the world faces, they require diplomatic solutions to implement them. How then can the promise of science be more effectively harnessed by the world of diplomacy?

For one thing, scientists could — and should — insert themselves more frequently into the diplomatic process to directly influence governments (Fedoroff, 2009; Lord and Turekian, 2007). Scientists are treated with a level of trust that other officials do not receive (Kaplan, 2011). In this connection, Jeffrey Boutwell (Chapter 10) makes the case that individual scientists can have a definite influence on the outcomes and decisions of governments. It was the scientists such as Joseph Rotblat involved in the production of nuclear weapons, and their analyses of the effectiveness of such weapons, that essentially helped to bring an end to the unrestrained arms race of the Cold War years. Similarly, it is pressure from scientists (along with campaigns spearheaded by Non-Governmental Organisations like *Greenpeace*) that is forcing governments to consider an internationally governed area in the Artic — comparable to the one that operates in the Antarctic — in order to prevent exploitation of resources and to protect the ecology of this area.

In addition, it is high time that United Nations — the chief custodian of the international interest — was reformed to ensure the organization could be a more effective forum for bringing diplomacy and science together to address global problems that can only be solved on a multilateral basis. Essentially, the current UN structure reflects the thinking of its founders in 1945 and assumes that the organization should serve the needs of sovereign states, particularly the most powerful ones in the early Cold War period. As an upshot, the permanent membership of the UN Security Council has remained virtually unchanged for more than 60 years and the permanent members hold the power to impose a veto on the Council's resolutions. This means the permanent five (P5) group not only sustains a structure that is no longer representative of today's political realities but also that the use of the veto for reasons of national interest effectively ensures that some of the world's most pressing problems are ignored or inadequately addressed. But if the doctrine of unfettered state sovereignty remains one of the biggest obstacles to a more effective UN, it is also evident that in an increasingly interconnected world key challenges require multilateral solutions, embracing science and diplomacy, and the use of the veto has become an encumbrance in this process.

Moreover, as Boutwell attests, social media and technology (read: the internet), which allows instantaneous communications between individuals that transcend borders, are also forces to be reckoned with in diplomacy. Science and technology are now inserting themselves into diplomacy directly. For example, in recent times Twitter has been used effectively to co-ordinate protests following the 2009 Iranian presidential election, to mobilise people during the Arab Spring, and to generate multifaceted opposition to genetically modified foods (Bennett *et al.*, 2014; Bruns *et al.*, 2013). States can no longer reckon with just the influence of their own people on policy: they are also exposed and susceptible to the views of the international community at levels that go well beyond traditional diplomacy and right down to the grassroots of citizen action. Put another way, diplomacy has always occurred in a metaphorical cloud far above the daily lives of mere citizens — but now it is influenced directly by a technological 'cloud' whereby the actions of governments are under scrutiny and, hence, influenced by the opinions and actions of citizens in a global context.

Conclusion

In conclusion, then, the sheer involvement of science in diplomacy and the inescapable need for science diplomacy to be integral to our futures in an increasingly science and technologically determined world, means that we can rule out the notion of science diplomacy being a 'false dawn'. As the weight of evidence in this book shows, science diplomacy is both real and here to stay. But there is an inherent optimism in characterising something as a 'new day': in this case, a promise that science diplomacy will provide solutions to problems that have hitherto proved intractable for normal diplomacy and political relationships. This is a promise that is largely unfulfilled by science diplomacy as of yet. Hence, rather than characterising the presence of science diplomacy as a 'new day' on the political landscape, it is more appropriate to think of it as 'early days'.

For science diplomacy to have real influence in the future, two things need to occur. In the short-term, there needs to be better communication of science to both diplomats and the populace. Ultimately, there needs to be an acceptance by nation states that some issues, at least, are so global in their reach and consequences that states need to forgo their perceive self-interest for the common good.

We do not yet know if science diplomacy will fulfil its promise, and so we do not yet know if this is the dawning of a new era in the way diplomacy is conducted internationally. But given the state of the planet and the issues facing us, we all had better hope that it does.

References

Ainley, D.G., Brooks C.M., Eastman J.T., and Massaro M. (2012) 'Unnatural Selection of Antarctic Toothfish in the Ross Sea, Antarctica', in Huettman, F. (ed.) *Protection of the Three Poles*, Tokyo: Springer Japan, pp. 53–75.

Backstrand, K. and Elgstrom O. (2013) 'The EU's role in climate change negotiations: From leader to "leadiator"', *Journal of European Public Policy*, 20(10): 1369–1386. DOI: 10.1080/13501763.2013.781781.

Bennett, W.L., Segerberg A., and Walker S. (2014) 'Organization in the crowd: Peer production in large-scale networked protests', *Information Communication and Society*, 17(2): 232–260. DOI: 10.1080/1369118X.2013.870379.

Bruns, A., Highfield T., and Burgess J. (2013) 'The Arab Spring and social media audiences: English and Arabic twitter users and their networks', *American Behavioral Scientist*, 57(7): 871–898. DOI: 10.1177/0002764213479374.

Buhr, K., Roth S., and Stigson P. (2014) 'Climate change politics through a global pledge-and-review regime: Positions among negotiators and stakeholders', *Sustainability*, 6(2): 794–811. DOI: 10.3390/su6020794.

Chatterton, P., Featherstone D., and Routledge P. (2013) 'Articulating Climate Justice in Copenhagen: Antagonism, the Commons, and Solidarity', *Antipode*, 45(3): 602–620. DOI: 10.1111/j.1467-8330.2012.01025.x.

den Elzen, M.G.J., Beltran A.M., Hof A.F., van Ruijven B. and van Vliet J. (2013) 'Reduction targets and abatement costs of developing countries resulting from global and developed countries' reduction targets by 2050'. *Mitigation and*

Adaptation Strategies For Global Change, 18(4): 491–512. DOI: 10.1007/s11027-012-9371-9.

Fedoroff, N.V. (2009) 'Science Diplomacy in the 21st Century', *CELL*, 136(1), 9-11. DOI: 10.1016/j.cell.2008.12.030.

Glomsrod, S., Wei T., and Alfsen K.H. (2013) 'Pledges for climate mitigation: the effects of the Copenhagen accord on CO_2 emissions and mitigation costs', *Mitigation and Adaptation Strategies For Global Change*, 18(5): 619–636. DOI: 10.1007/s11027-012-9371-9,

Greenwald, G., MacAskill E. and Poitras L. (2013) 'Edward Snowden: the whistleblower behind the NSA surveillance revelations', 10 June, *The Guardian*, Available: http://www.theguardian.com/world/2013/jun/09/edward-snowden-nsa-whistleblower-surveillance

Hallding, K., Juerisoo M., Carson M., and Atteridge A. (2013) 'Rising powers: the evolving role of BASIC countries', *Climate Policy*, 13(5): 608–631. DOI: 10.1080/14693062.2013.822654.

Hardin, G. (1968) 'The Tragedy of the Commons', *Science*, 162(3859): 1243–1248. DOI: 10.1126/science.162.3859.1243.

Held, D. (2010) *Cosmopolitanism: Ideals and Realities*, Cambridge: Polity Press, pp. 1–26.

Hof, A.F., den Elzen M.G.J. and Roelfsema M. (2013) 'The effect of updated pledges and business-as-usual projections, and new agreed rules on expected global greenhouse gas emissions in 2020', *Environmental Science and Policy*, 33: 308–19. DOI: 10.1016/j.envsci.2013.06.007.

Hope, C. and Hope M. (2013) 'The social cost of CO2 in a low-growth world', *Nature Climate Change*, 3: 722–724. DOI: 10.1038/nclimate1935.

Jiang, K., Zhuang X., Miao R. and He C. (2013) 'China's role in attaining the global 2°C target', *Climate Policy*, 13: 55—9.

Kaplan, K. (2011) 'International opportunities: The science of diplomacy', *NatureJobs*, 470(7334): 425–427. DOI: 10.1038/nj7334-425a.

Koenig, R. (2009). US Takes Steps to Use Science To Improve Ties to Muslim World. *Science*, 326(5955): 920–921. DOI: 10.1126/science.326.5955.920.

Lord, K.M. and Turekian V.C. (2007) 'Time for a new era of science diplomacy', *Science*, 315(5813): 769–770. DOI: 10.1126/science.1139880.

Luderer, G., Pietzcker R.C., Bertram C., Kriegler E., Meinshausen M. and Edenhofer O. (2013) 'Economic mitigation challenges: how further delay closes the door for achieving climate targets', *Environmental Research Letters*, 8(3): 034033. DOI: 10.1088/1748-9326/8/3/034033.

Neslen, A. (2013) 'Russian 'fishing interests' threaten bid for Antarctic conservation area', 16 July, *EurActiv*, Available: http://www.euractiv.com/science-policymaking/vote-world-largest-protected-mar-news-529309.

Oels, A. (2013) 'Rendering climate change governable by risk: From probability to contingency', *Geoforum*, 45: 17–29. DOI: 10.1016/j.geoforum.2011.09.007.

Rampton, R (2014) 'Obama to propose ending NSA bulk collection of phone records: Official', 25 March, *Reuters*, Available: http://www.reuters.com/article/2014/03/25/us-usa-security-obama-nsa-idUSBREA2O03O20140325.

Rogelj, J., McCollum D.L., O'Neill B.C. and Riahi K. (2013) '2020 emissions levels required to limit warming to below 2°C'. *Nature Climate Change*, 3(4): 405–412. DOI: 10.1038/nclimate1758.

Sarkadi, B. and Schatten G. (2012) 'Stem Cell Course in the Middle East: Science Diplomacy and International Collaborations During the Arab Spring', *Stem Cell Reviews and Reports*, 8(1): 87–90. DOI: 10.1007/s12015-011-9277-z.

Shum, R., Y. (2014) 'China, the United States, bargaining, and climate change', *International Environmental Agreements-Politics Law and Economics*, 14(1): 83–100. DOI: 10.1007/s10784-013-9231-4.

The Daily Telegraph (2010) 'Australia's ban on New Zealand apples overturned by World Trade Organisation—report', 12 April, *The Daily Telegraph*, Available: http://www.dailytelegraph.com.au/archive/news/australias-ban-on-new-zealand-apples-overturned-by-world-trade-organisation-report/story-e6frez7r-1225852857907.

Index